ENVIRONMENTAL ISSUES IN PACIFIC NORTHWEST FOREST MANAGEMENT

Committee on Environmental Issues
in Pacific Northwest Forest Management

Board on Biology

Commission on Life Sciences

National Research Council

NATIONAL ACADEMY PRESS
Washington, D.C.

NATIONAL ACADEMY PRESS 2101 Constitution Ave., N.W. Washington, D.C. 20418

NOTICE: The project that is the subject of this report was approved by the Governing Board of the National Research Council, whose members are drawn from the councils of the National Academy of Sciences, the National Academy of Engineering, and the Institute of Medicine. The members of the committee responsible for the report were chosen for their special competences and with regard for appropriate balance.

This project was supported by Contract No. 53-3187-3-02 between the National Academy of Sciences and the U.S. Department of Agriculture, by Grant No. 955-0585 from the Ford Foundation, and by funds from the National Research Council. Any opinions, findings, conclusions, or recommendations expressed in this publication are those of the author(s) and do not necessarily reflect the view of the organizations or agencies that provided support for this project.

Library of Congress Catalog Card Number 00-106115
International Standard Book Number 0-309-05328-5

Photograph courtesy of the Forest History Society, Durham, N.C.

Additional copies of this report are available from:

National Academy Press
2101 Constitution Ave., NW
Box 285
Washington, DC 20055

800-624-6242
202-334-3313 (in the Washington metropolitan area)
http://www.nap.edu

Copyright 2000 by the National Academy of Sciences. All rights reserved.

Printed in the United States of America

THE NATIONAL ACADEMIES

National Academy of Sciences
National Academy of Engineering
Institute of Medicine
National Research Council

The **National Academy of Sciences** is a private, nonprofit, self-perpetuating society of distinguished scholars engaged in scientific and engineering research, dedicated to the furtherance of science and technology and to their use for the general welfare. Upon the authority of the charter granted to it by the Congress in 1863, the Academy has a mandate that requires it to advise the federal government on scientific and technical matters. Dr. Bruce M. Alberts is president of the National Academy of Sciences.

The **National Academy of Engineering** was established in 1964, under the charter of the National Academy of Sciences, as a parallel organization of outstanding engineers. It is autonomous in its administration and in the selection of its members, sharing with the National Academy of Sciences the responsibility for advising the federal government. The National Academy of Engineering also sponsors engineering programs aimed at meeting national needs, encourages education and research, and recognizes the superior achievements of engineers. Dr. William A. Wulf is president of the National Academy of Engineering.

The **Institute of Medicine** was established in 1970 by the National Academy of Sciences to secure the services of eminent members of appropriate professions in the examination of policy matters pertaining to the health of the public. The Institute acts under the responsibility given to the National Academy of Sciences by its congressional charter to be an adviser to the federal government and, upon its own initiative, to identify issues of medical care, research, and education. Dr. Kenneth I. Shine is president of the Institute of Medicine.

The **National Research Council** was organized by the National Academy of Sciences in 1916 to associate the broad community of science and technology with the Academy's purposes of furthering knowledge and advising the federal government. Functioning in accordance with general policies determined by the Academy, the Council has become the principal operating agency of both the National Academy of Sciences and the National Academy of Engineering in providing services to the government, the public, and the scientific and engineering communities. The Council is administered jointly by both Academies and the Institute of Medicine. Dr. Bruce M. Alberts and Dr. William A. Wulf are chairman and vice chairman, respectively, of the National Research Council.

COMMITTEE ON ENVIRONMENTAL ISSUES IN
PACIFIC NORTHWEST FOREST MANAGEMENT

NORMAN L. CHRISTENSEN, JR. (Chair), Duke University, Durham, North Carolina
STANLEY V. GREGORY, Oregon State University, Corvallis, Oregon
PERRY R. HAGENSTEIN, Wayland, Massachusetts
THOMAS A. HEBERLEIN, University of Wisconsin, Madison, Wisconsin
JOHN C. HENDEE, University of Idaho, Moscow, Idaho
JEFFREY T. OLSON, Peekskill, New York
JAMES M. PEEK, University of Idaho, Moscow, Idaho
DAVID A. PERRY, Oregon State University, Corvallis, Oregon
TIMOTHY D. SCHOWALTER, Oregon State University, Corvallis, Oregon
KATHLEEN SULLIVAN, Sustainable Ecosystems Institute, Portland, Oregon
G. DAVID TILMAN, University of Minnesota, St. Paul, Minnesota
KRISTINA A. VOGT, Yale University, New Haven, Connecticut

Staff

LEE R. PAULSON, Project Director
PATRICIA PEACOCK, Program Officer
ERIC FISCHER, Principal Staff Officer
KATHRINE IVERSON, Project Assistant
MIRSADA KARALIC-LONCAREVIC, Information Specialist
STEPHANIE PARKER, Graphics and Layout

BOARD ON BIOLOGY

PAUL BERG, Chairman, Stanford University School of Medicine, Stanford, CA
JOANNA BURGER, Rutgers University, Piscataway, NJ
MICHAEL T. CLEGG, University of California, Riverside, CA
DAVID EISENBERG, University of California, Los Angeles, CA
DAVID J. GALAS, Keck Graduate Institute of Applied Life Science, Claremont, CA
DAVID V. GOEDDEL, Tularik, Inc., South San Francisco, CA
ARTURO GOMEZ-POMPA, University of California, Riverside, CA
COREY S. GOODMAN, University of California, Berkeley, CA
CYNTHIA K. KENYON, University of California, San Francisco, CA
BRUCE R. LEVIN, Emory University, Atlanta, GA
ELLIOT M. MEYEROWITZ, California Institute of Technology, Pasadena, CA
ROBERT T. PAINE, University of Washington, Seattle, WA
RONALD R. SEDEROFF, North Carolina StateUniversity, Raleigh, NC
ROBERT R. SOKAL, State University of New York, Stony Brook, NY
SHIRLEY M. TILGHMAN, Princeton University, Princeton, NJ
RAYMOND L. WHITE, University of Utah, Salt LakeCity, UT

RALPH DELL, Executive Director

COMMISSION ON LIFE SCIENCES

MICHAEL T. CLEGG *(Chair)*, University of California, Riverside, CA
PAUL BERG *(Vice Chair)*, Stanford University, Stanford, CA
FREDERICK R. ANDERSON, Cadwalader, Wickersham & Taft, Washington, DC
JOANNA BURGER, Rutgers University, Piscataway, NJ
JAMES E. CLEAVER, University of California, San Francisco, CA
DAVID EISENBERG, University of California, Los Angeles, CA
JOHN EMMERSON, Fishers, IN
NEAL FIRST, University of Wisconsin, Madison, WI
DAVID J. GALAS, Keck Graduate Institute of Applied Life Science, Claremont, CA
DAVID V. GOEDDEL, Tularik, Inc., South San Francisco, CA
ARTURO GOMEZ-POMPA, University of California, Riverside, CA
COREY S. GOODMAN, University of California, Berkeley, CA
JON W. GORDON, Mount Sinai School of Medicine, New York, NY
DAVID G. HOEL, Medical University of South Carolina, Charleston, SC
BARBARA S. HULKA, University of North Carolina, Chapel Hill, NC
CYNTHIA KENYON, University of California, San Francisco, CA
BRUCE R. LEVIN, Emory University, Atlanta, GA
DAVID LIVINGSTON, Dana-Farber Cancer Institute, Boston, MA
DONALD R. MATTISON, March of Dimes, White Plains, NY
ELLIOT M. MEYEROWITZ, California Institute of Technology, Pasadena, CA
ROBERT T. PAINE, University of Washington, Seattle, WA
RONALD R. SEDEROFF, North Carolina State University, Raleigh, NC
ROBERT R. SOKAL, State University of New York, Stony Brook, NY
CHARLES F. STEVENS, The Salk Institute, La Jolla, CA
SHIRLEY M. TILGHMAN, Princeton University, Princeton, New Jersey
RAYMOND L. WHITE, University of Utah, Salt Lake City, UT

Staff

WARREN R. MUIR, Executive Director
JACQUELINE K. PRINCE, Financial Officer
BARBARA B. SMITH, Administrative Associate
LAURA T. HOLLIDAY, Senior Program Assistant

PREFACE

In response to a congressional request, the Board on Biology convened the Committee on Environmental Issues in Pacific Northwest Forest Management in 1993 to review information concerning the status of resources of the Pacific Northwest and the relationship of those to supply and demand for forest products in other regions of the country. Committee members were selected for their expertise in forestry practices, public lands issues, biology, vertebrate and invertebrate ecology, rural sociology, and multiple-use land management.

The committee grappled with many contentious questions. What activities ought to be included under the rubric of "forest management?" What was the presettlement character of Pacific Northwest forest landscapes and how have those forests and landscapes been altered by human activities? What are the ecological and economic consequences of changes in forest management practices? How are changes in forest management policy and practice in the Pacific Northwest influencing forest management outside that region? The definition and assessment of old-growth forests and their associated biodiversity and the implications of changing modes of management for their future were a particularly important part of the committee's deliberations. Nine meetings with guest presentations were held, including meetings in Portland, Oregon; Seattle, Washington; and Post Falls, Idaho. The public was invited to briefings in Washington, D.C., and Portland, Oregon.

The original request for this study came from Congress in 1992, and much has changed in between then and now. There is now little debate over the value and importance of conserving important areas of old growth forest. Conflict over forest-resource management continues, of course, but it is no longer polarized along such simple axes as "owls versus jobs." However, the basic question of how we are to achieve sustainable management of Pacific Northwest forests, as well as the forests of other regions in the context of increasing demand for the goods and services forests provide remains. How such specific initiatives as the Northwest Forest Plan fit into that objective is also unclear. It was clear throughout our study that few institutions exist to resolve the increasingly complex conflicts between the needs and aspirations of local communities and regional visions for the health of forest ecosystems. Improvement in communication, institutional learning, and institutional performance is badly needed among government agencies, environmental groups, the business sector and the academic world.

The problems presented in our charge are highly interdisciplinary, ranging from such fundamental natural science problems as the presettlement dynamics of forests and the implications of forest fragmentation for the management of species populations to very human impacts of changes in forest management on regional economies and rural communities. Furthermore, the committee was constantly aware of the reality that forest management in all regions is being undertaken with an increasingly diverse array of objectives across an increasingly complex array of ownerships. When the committee began its work, FEMAT (1993) had just been published and the President's Northwest Forest Plan just released. Over these several years, a number of the management challenges and our understanding of a number of particular issues have changed. We have worked hard to incorporate those changes into this document. I am grateful to the committee for their willingness, indeed eagerness, to work across traditional disciplinary lines and perseverance in a process that became more drawn out than any of us expected.

The committee would like to acknowledge the generous assistance of numerous persons who made presentations or in other ways enlightened the committee regarding the complex issues facing the Pacific Northwest, including Senator Mark Hatfield, U.S. Senate; Mark Walker, staff to Senator Hatfield; Jack Ward Thomas, U.S. Forest Service;

David Darr, U.S. Forest Service; Nancy Foster, National Marine Fisheries Services; Donal Knowles and Donald Barry, U.S. Department of the Interior; Stuart McKenzie, U.S. Geological Survey; Mike Penfold, Bureau of Land Management; James E. Brown, Oregon Department of Forestry; Kaleen Cottingham, State of Wahsington; Robert Ewing, California Department of Forestry and Fire Protection; Jay O'Laughlin, University of Idaho; Charley Grenier, Plum Creek, Columbia Falls, Montana; Kevin Bolling, Potlach Corporation, Lewiston, Idaho; Ted Strong, Columbia River Inter-Tribal Commission, Portland, Oregon; Si Whitman, Nez Perce Tribe, Lapwai, Idaho; Gary Morishima, Intertribal Timber Council, Portland, Oregon; Terry Williams, Northwest Indian Fisheries Commission, Olympia, Washington; Jack Shipley and Jim Neal, Applegate Partnership, Grants Pass, Oregon; and Leah Wills and Jim Wilcox, Plumas Corporation, Quincy, California. A grant from the Ford Foundation provided important support to complete this project.

This report has been reviewed in draft form by individuals chosen for their diverse perspectives and technical expertise in accordance with procedures for reviewing NRC reports approved by the NRC's Report Review Committee. The purpose of this independent review is to provide candid and critical comments that assist the NRC in making the published report as sound as possible and to ensure that the report meets institutional standards for objectivity, evidence, and responsiveness to the study charge. The content of the final report is the responsibility of the NRC and the study committee and not the responsibility of the reviewers. The review comments and draft manuscript remain confidential to protect the integrity of the deliberative process. We wish to thank the following individuals, who are neither officials nor employees of the NRC, for their participation in the review of this report: Ellis Cowling, North Carolina State University; Paul Ellefson, University of Minnesota; Jerry Franklin, University of Washington; and John Lattin, Oregon State University.

The individuals listed above have provided many constructive comments and suggestions. Nevertheless, the responsibility for the final content of this report rests entirely with the authoring committee and the NRC.

Several NRC staff members contributed in important ways to the committee's deliberations and to the assembly of this final report. Patricia Peacock and Eric Fischer each played important roles in our early discussions and the assembly of many early report draft reports. We also thank Gordon Orians for his thoughtful review of and comments on the draft report that was sent to review. We are especially grateful to Lee Paulson whose writing, editorial, diplomatic, and motivational skills contributed mightily to this final product.

Norman Christensen, Chair

Contents

Executive Summary   1

1   *The Promised Land: The Land of Promise*   15
Introduction, 15
   Loss of the Frontier, 16
   Changing Knowledge Base, 16
   Changing Social Values, 17
The Study Area, 21
Other Major Studies and Reports, 22
This Report, 26

2   *The Region and Its Forests*   27
Introduction, 27
A Brief History, 27
Demographics and the Economy, 33
   Population Growth, 33
   Economic Implications of Population Growth, 35
The Region's Forests, 36
   Westside Forests, 36
   Eastside Forests, 38
   Northern Rocky Mountains, 39
Regional Ownership Patterns, 40
Summary, 42

3   Old-Growth Forests   *44*

Introduction, 44
What Is Vegetative Succsession?, 44
What Is an Old-Growth Forest?, 45
   Time Required for Old-Growth Development, 53
   Old-Growth Landscapes, 53
   Managed Forests and Old-Growth Characteristics, 59
   Stand Size, 60
Biological Functions of Old-Growth and Late-Successional
   Forests, 61
   Species Diversity, 62
   Logs and Woody Debris, 64
   Susceptibility to Disturbance, 65
Aquatic Ecosystems, 66
Extent and Status of Old-Growth Forests, 67
Summary, 72

**4   The Status and Functioning of Pacific Northwest
Forests**   *73*

Introduction, 73
Forest Condition, 73
   General Criteria of Condition, 73
   The Role of Biological Diversity, 75
   Resistance and Resilience, 78
   Landscape Change and Threats to Biodiversity, 81
   Diseases and Pests, 85
   Incidence of Pest and Disease Outbreaks, 87
Status of Other Plant Species, 91
Status of Wildlife, 92
   Salmon and Other Fisheries, 98
   Invertebrates, 101
   Fungi, 102
Viable Populations and the Conservation of Biodiversity, 103
   Demographic Factors, 105
   Genetic Factors, 106
   Population Viability Analysis, 107

5   Forest Succession, Fire, and Landscape Dynamics   *108*

Introduction, 108
Concepts of Succession and Landscape Dynamics, 109
Fire and Landscape Dynamics, 112
   Presettlement Fire Regimes and Successional Change, 113
   Human Alteration of Fire Regimes, 116
   Natural Disturbance and Human Management: An Ecological Comparison, 118
      Landscape Considerations, 118
      Fuels, 119
      Nutrient Fluxes, 120
      Biological Diversity, 121

6   Products from the Forests   122

Pacific Northwest Wood Products in the National Economy, 123
Products in National and International Markets, 129
Effects of Changes in Federal Timber Harvests in the Pacific Northwest, 129
   Increased Harvests at the Extensive Margin, 135
   Increased Harvests at the Intensive Margin, 136
   Increased Use of Hardwoods, 137
   Technology Changes, 138
   Materials Substitution, 138
Environmental Effects, 139
Implications for Other Regions, 141
Effects on Regional and National Income, 144
Change and Incentives, 145
Nonwood Products from Forests, 146
   Wildlife-Related Recreation, 147
   Other Forest-Related Recreation, 151
   Fisheries, 153
   Mushrooms, 155
   Water, 156
   Effects of Changes in Management of PNW Forests on Nonwood Products, 157
Regional Economic Effects, 158
Summary, 159

7   **Forest Management and Rural Communities in the
      Pacific Northwest**   160
    Rural Economic Well Being and Natural-Resource Industries, 160
    Timber Dependency and Community Well Being, 162
    Diversification in Rural Communities of the Pacific Northwest, 165
    In-Migration, 168
    Poverty and Plenty — Anger and Hope in the Pacific Northwest, 169
    Summary, 170

8   **A Framework for Sustainable Forest Management**   171
    Introduction, 171
    Elements of Forest Management, 171
        Land Allocation, 172
        Rationing Uses, 174
        Harvesting, 176
        Investment, 179
            Examples of Forest Investment, 180
            Positive and Negative Incentives to Forest Investment, 181
        Meeting the Goals of Sustainable Forest Management, 182
    A Framework for Sustainable Forest Management, 184
        Operational Goals, 184
        Managing in Context and Across Scales of Space and Time, 185
    Approaches to Managing for Diversity: General Considerations, 186
        Reserves, 189
        Logging to Improve Forest Health, 190
        Management for Complexity and Diversity, 191
        Variability and Change, 192
        Uncertainty and Surprise, 193
        Humans as Ecosystem Components, 194
        Making Management Adaptable, 195
        Resolving Conflicts, 197

9   **Conclusions and Recommendations**   199
        Forestry Practices in the Pacific Northwest, 200
        What Is Old-Growth?, 205
        Old-Growth Management, 206
        Forest Products Substitution, 206
        Research Recommendations, 208

References	*211*
Glossary	*254*

ENVIRONMENTAL ISSUES IN PACIFIC NORTHWEST FOREST MANAGEMENT

EXECUTIVE SUMMARY

In his 1938 dust-bowl-era treatise, Richard L. Neuberger described the Pacific Northwest as "the promised land," and certainly the region has appeared to waves of immigrants as a land of great bounty and promise. The pioneers who followed the Oregon Trail were attracted by the wealth of natural resources–forests, fisheries, rivers, fertile soils–and the opportunity for a new future. Later immigrants were attracted for much the same reasons, but development of the region's natural resources has led to problems. Among them has been the depletion of old-growth forests, which had supported the development of the wood products industry.

With a growing recognition of the importance of old-growth forests for sustaining biodiversity and species dependent on old growth, conflicts with logging increased during the 1980 and 1990s. Intervention by President Clinton in 1993 led to adoption of the Northwest Forest Plan. This brought about a sharp reduction in federal timber harvests, the establishment of substantial areas of old-growth reserves, and adoption of environmentally sensitive forest practices on federal forests.

Congress asked the National Academy of Sciences to provide basic information to help guide future forest management in the region, and the National Research Council convened the Committee on Environmental Issues in Pacific Northwest Forest Management in 1993. The committee was charged to

• Review the information concerning the current state of knowledge of forest resources.

- Review definitions of old growth, including biological, economic, social, and physical amenities that old-growth forest provides; provide analyses of age at which a forest becomes old growth; and evaluate amenities characteristic of old-growth forests that can be preserved under different harvesting regimes.
- Review forest management practices and the effects on resources of the forests and the economic consequences of those practices.
- Review the use of forest products from the Pacific Northwest and the degrees to which forest products from other parts of the United States can be substituted for them.

Clear goals are essential to any effort to rationalize forest management in the Pacific Northwest. The committee identified four goals that it believes are at the heart of the issues in forest management in this region (see Chapter 8):

- Sustain viable populations of indigenous species
- Maintain properly functioning ecological processes
- Meet human needs for forest commodities
- Satisfy cultural and aesthetic values

Those goals provided the general framework for the committee's study and helped to clarify the kinds of issues that are involved in Pacific Northwest forest management. Conflicts arise among them as they are applied in specific circumstances because they cannot all be maximized or optimized simultaneously. Much of the disagreement regarding policies and protocols in forest management is associated with the relative priorities that should be assigned to each of those goals.

THE REGION

Significant intraregional variation exists among Pacific Northwest forests. West of the crest of the Cascade Mountains (the "Westside") where rainfall is high, forests are among the most massive in the world. Dominant trees include Douglas-fir, several species of true fir, western red cedar, Sitka spruce, and western hemlock. In the moist conditions, fires are infrequent and fire-return intervals relatively long and highly variable. Drier conditions prevail east of the Cascades in Washington and Oregon (the "Eastside") with grasslands and shrublands grading into ponderosa pine forests at lower elevations and lodgepole pine dominating higher elevation sites. Historically, fires in this subregion were more frequent, and fire exclusion in recent decades has encouraged ingrowth of shrubs and shade tolerant trees. The Northern Rocky Mountain subregion of northeastern Washington, northern Idaho, and western Montana is complex with regard to climate, substrate, and biota. Forests in this subregion are influenced by catastrophic, stand-destroying wildfires. Douglas-fir and true firs dominate some sites, but lodgepole and ponderosa pines, as well as larch, are common. Quaking aspen is widely distributed across this region in burned and logged areas.

Basic patterns of land ownership were set in the mid-nineteenth century. Federal public-domain lands were sold or granted to encourage settlement and development; others were established as national parks and forests. Substantial lands went directly into private ownership under various homestead and other laws. Grants to railroads in the region were made in broad swatches, creating a well-known "checkerboard" ownership pattern that has complicated management.

About half the land in the Pacific Northwest is in public ownership; most of that land is managed by the U.S. Forest Service (USFS) and the Bureau of Land Management (BLM). The states have forest and other natural resource lands, most of which were obtained from the federal government when statehood was granted. American Indian tribes and the Bureau of Indian Affairs manage large tracts of land on reservations scattered throughout the region. Private owners have holdings that range from small woodlots to millions of acres of industrial forests owned by large corporations.

THE DYNAMICS OF PACIFIC NORTHWEST LANDSCAPES

Disturbances such as fire, wind, and insect and pathogen outbreaks occur naturally in all Pacific Northwest forest types, although the frequency, intensity, and spatial extent of such disturbances vary considerably. Such variations generate different patterns of forest succession and thereby contribute to the diversity of Pacific Northwest landscapes. Among the various kinds of natural disturbance, fire has been most important. On average, fire frequency decreases with increasing moisture availability, and fire intensity is generally inversely related to frequency. In forests typified by short fire-return intervals, fires are usually confined to the fine fuels on the forest floors. In addition to permitting natural succession of other species, longer fire-free intervals permit invasion of transgressive trees and accumulation of fuels that can carry fires into the forest canopy.

Before settlement of the Pacific Northwest, fire-return intervals and fire intensities varied considerably among the forest types of the Westside. Moderate-severity fires at 25-100 year intervals were the norm for Douglas-fir (without hemlock), mixed evergreen, red fir, and lodgepole pine forests. Longer fire-free intervals (100 to 400 years) and more intense fires were typical of other Westside forest types. The drier ponderosa pine forests of the Eastside typically experienced frequent (5-10 year) fires, which kept understories open and minimized fuel accumulations.

To the extent possible and within the range of natural variability, disturbances such as fire should be maintained. Fire regimes have been altered in complex ways by human activities. Development of urban centers and transportation networks, increased recreational use of forests, and timber activities have increased the frequency of fire starts. Suppression of light surface fires has resulted in increased accumulation of fuels in many Pacific Northwest forests. That has been especially important in the ponderosa pine forests of the Eastside and in Idaho and Montana. The extent and landscape-level continuity of early-successional forests has increased across the region. Taken together, those factors have resulted in conditions that are favorable for extensive, high-intensity fires in these parts of the region.

Although prescribed fire is important and necessary to reduce fuel accumulations and restore landscapes to a less flammable condition, it is not possible to deal with the spatial extent of the problem with that tool alone. Thus, logging and other human management interventions have been proposed as surrogates for fire in this regard. Just as the effects on ecological processes and biological diversity of natural disturbances such as fire are highly variable, so are the effects of human activities, such as logging and forest thinning. Without attention to specific goals such as fuel loads, structural features or "legacies" important to post-disturbance regeneration, biodiversity, and patterns of post-harvest recovery, silvicultural treatments might not simulate important fire effects nor have the desired effect on landscape flammability. More research is needed into the importance of specific factors.

Recommendation: The important roles of natural disturbances and legacies in sustaining ecological processes must be recognized in forest-management practices for both federal and nonfederal forests in the Pacific Northwest.

THE BIOLOGICAL DIVERSITY OF PACIFIC NORTHWEST FORESTS

Much of the biological diversity of Pacific Northwest is associated with late-successional and old-growth forests. If stable and sustainable forests and associated ecosystem amenities are desired in the Pacific Northwest, those forests should be managed to preserve genetic and species diversity. The remaining late-successional and old-growth forests could form the cores of regional forests managed for truly and indefinitely sustainable production of timber, fish, clean water, recreation, and numerous other amenities of forested ecosystems. The multiple threats to biodiversity caused by past forestry practices and the effects of lost biodiversity on ecosystem processes and sustainability demand a new approach to forest management that acknowledges the connection between biodiversity and the sustained provision of commodities and amenities from forests.

Further cutting of the remaining late successional and old-growth forests of the Pacific Northwest is expected to cause rapidly accelerating threats to the biological diversity of the region. The number of fish,

> *The multiple threats to biodiversity caused by past forestry practices and the effects of lost biodiversity on ecosystem processes and sustainability demand a new approach to forest management....*

bird, mammal, amphibian, plant, and invertebrate species already threatened or endangered with extinction because of past land-use practices likely represent about one-fourth of the number of species that would be threatened if just half of the remaining late-successional and old growth forests on public lands were to be harvested. If preservation of biodiversity is to be effective, it is imperative that the region be managed to maintain within the landscape at least the current proportion of late-successional and old-growth forests. These should be supplemented with adjacent or nearby second-growth forests, including naturally regenerated stands, that are allowed to attain old-growth characteristics by having rotation times of at least 150 years. In so doing, it is critical that the ecosystem types that have received greater proportional cutting, especially the low-elevation forests of the Westside, be provided the highest level of protection and restoration. Research is needed to develop restoration prescriptions. Forest vitality should be restored and maintained. Development of a clear strategy is crucial if biodiversity is to be protected; management by intuition is insufficient. Success of such a strategy will depend on preserves, and success of preserves will depend on management of the surrounding landscape.

> **Recommendation:** Forest management in the Pacific Northwest should include the conservation and protection of most or all of the remaining late-successional and old-growth forests. Protected areas that include late-successional and old-growth forests should have an important role in an overall strategy for forest management in the region.

Not all ecosystem types are represented on public lands in the Pacific Northwest–examples of gaps in coverage include lowland floodplain forests, oak woodlands, and coastal tidal marshes. Checkerboard ownership of public and private lands hinders effective management of forest ecosystem patterns and processes. Opportunities such as land exchanges might offer ways to obtain critical habitats and create public and private management boundaries that are consistent with the behavior of ecosystem processes.

Executive Summary

> Recommendation: Goals for protected late-successional and old-growth reserves should include representation of the range of forested ecosystems in the region. This should include rationalization of reserve boundaries, and land exchanges between public and private landowners should be pursued.

OLD-GROWTH FORESTS

Old-growth forests are defined by five characteristics:

1. A minimum density (16-50/ha) of large (52-92 cm diameter at breast height (dbh)) trees (the exact number and size thresholds to qualify as old growth vary among forest types)
2. High standard deviation in tree diameters
3. Tree decadence, i.e., broken tree tops, excavated bole cavities, root-collar cavities, and bark resinosis
4. Presence of large dead wood
5. Complex, multilayered tree canopies

Old-growth forests are ecologically unique with respect to their complexity and biodiversity, accumulations of logs and woody debris, and resistance to disturbances, such as fire and insect and fungal pest outbreaks. But there is no precise threshold age at which forests become old-growth. Forests can be classified as old growth as early as 150 years or as late as 250 years. Forests acquire old-growth characteristics gradually, showing some earlier than others. The rates of acquisition of old-growth characteristics vary with nutrient and moisture availability and residual forest components from the predisturbance stand. It is important to acknowledge that old growth is a continuum of processes, rather than a simple definition. Some researchers have developed indexing approaches to accommodate that continuum.

In the western Cascades, old-growth forests have been reduced from 40-70% of the landscape in presettlement times to 13-18% today. Estimates of presettlement extent of old-growth forests vary depending on assumptions about the frequency of crown-killing fires. Old-growth ponderosa pine forests might have composed as much as 90% of the lower and middle elevations of the eastern Cascades. Approximately 20% of public and private Eastside lands remain in old-growth today. On Westside and Eastside landscapes, old-growth forest stands are, on average, becoming smaller and increasingly fragmented.

Late-successional forests, which are not the same as old-growth forests, are characterized by the invasion of shade-tolerant species. This usually occurs only in the absence of significant disturbances, which maintain the dominance of less shade-tolerant species, such as Douglas-fir and ponderosa pine. Thus, because of intermittent fires that tend to maintain these species even to old age, true late-successional forests are relatively rare in the Pacific Northwest.

FOREST PRODUCTS SUBSTITUTION

Increased timber harvests in the U. S. South and increased softwood lumber imports from Canada, both in response to ordinary market forces, together have offset the reduced timber harvests on federal forests in the West. Total consumption of softwood wood products in the United States does not appear to have been substantially reduced. The reduction in federal timber harvests has been accompanied by some increase in the price of softwood lumber to consumers and in more substantial increases in prices paid for timber that is harvested from federal and nonfederal forests.

The expected effects of adopting the Northwest Forest Plan on some biological resources in the Pacific Northwest were examined at length and were largely addressed in the plan (FEMAT 1993). The potential effects on products other than timber, such as recreation and special forest products, were not thoroughly evaluated in that report, partly because of the lack of good information. But it is clear that the reductions in federal timber harvests in the Pacific Northwest favor some kinds of game and nongame species of wildlife over others, affect hunting conditions, improve habitat for fisheries, and maintain opportunities for recreation in the region.

Sustaining the increased level of timber harvests in the South, which come mainly from private forests, will require more intensive management practices because of the reduction in federal timber harvests in the

West. The possible effects of such practices on biological resources, such as wetlands and the red cockaded woodpecker, apparently have not been carefully evaluated. Similarly, possible effects on employment and communities in the South have not been carefully evaluated.

Pressures on forests for all uses in the Pacific Northwest and elsewhere in the United States will probably continue to rise in response to basic demands for materials, space, and environmental amenities. The increasing production of wood products on private forests is leading to lower ages of trees at harvest and more intensive silvicultural operations such as thinning, use of improved genetic stock for single-species planting, fertilization, and increased use of pesticides. Tracking these changes and being prepared to take action when the effects are judged to be serious are challenges for public policy.

> **Recommendation: Regional assessments of the impacts of increasingly intensive forest-management practices, especially on private forests, should be conducted to evaluate the impacts of shifting regional patterns of timber harvesting. In particular, an assessment is needed of the effects on key species and ecosystems in the U.S. South of increased timber harvests and management intensity that has resulted from reduced timber harvests on federal forests in the West.**

FOREST MANAGEMENT AND HUMAN COMMUNITIES

Claims that adopting the Northwest Forest Plan would devastate the economy of the Pacific Northwest have proven to be greatly overstated. The impacts on the region's overall employment and income have been modest. More timber-related jobs have been lost in recent decades to increases in efficiency and productivity than to reductions in timber harvests. At the same time, some communities heavily dependent on federal timber harvests have had a difficult time, although even these impacts have been eased by the region's buoyant economy.

As with other kinds of rural communities dependent on extractive resources, rural communities in the region are experiencing the overall social consequences of changing resource use. The social forces that affect rural areas are powerful, but poorly understood. The conceptual, analytical, and information bases for relating changes in forest management to economic and social consequences need to be improved.

Experience with the multiyear process leading to adoption of the Northwest Forest Plan also suggests the need to address ways to reduce conflicts over forest management issues. Alternative mechanisms for dispute resolution need to be examined and their appropriateness for situations such as that in the Pacific Northwest evaluated.

> **Recommendation: Experience with FEMAT, the Northwest Forest Plan, and other processes used to help resolve disputes over Pacific Northwest forestry practices should be used to explore alternative mechanisms for dispute resolution.**

FOREST MANAGEMENT FOR THE FUTURE

Forest management is far more than logging, silviculture, and fiber extraction. It must account for management of a variety of landscapes to achieve maintenance of key processes. The committee viewed forest management within the context of four elements: allocation of land to particular uses, rationing or scheduling of use, harvest of forest products, and investment in productive resources. Pacific Northwest forest managers face significant challenges with regard to each of those elements:

Allocation. Most allocation decisions are made within particular ownerships and, thus, at scales smaller than ecosystems and landscapes.

Rationing uses. No single rotation age or age between successive cuts fits all forest management goals or circumstances. However, the age at which forests become "financially mature" (i.e., the age at which the cost of holding trees exceeds the increase in value expected over that time) from the standpoint of fiber production is often considerably shorter than optimal rotation ages for other management objectives.

Harvesting. Logging and postharvest planting methods have changed through time with changes in our knowledge and differences in owners' objectives. In general, harvest practices are changing in ways that maintain or increase site productivity that reduce impacts on forest values, such as biodiversity and aesthetics, but further improvements in this area are needed.

Investment. Positive and negative incentives are influencing forest-management investments in the Pacific Northwest. Higher timber prices provide financial incentives for investments in timber production, such as postharvest site treatment, fertilizer application, and precommercial and commercial thinning. However, uncertainty regarding interpretation of the Endangered Species Act may be creating incentives for premature timber harvest and reduced management levels on some private lands. This has been offset to some extent by some private forest owners who adopt habitat conservation plans, which remove some of the incentives for premature logging.

Ecosystem management or sustainable forest management provides a framework for forest management in the context of competing goals and objectives and across scales of time and space. Key elements of this framework are

Operational goals. Goals should be formulated in terms of ecosystem processes, as well as economic and social outcomes, so as to provide measurable benchmarks for success of management policies and practices.

Context and scale. Managers must be cognizant of connectivity within forest landscapes and recognize that activities at one location influence processes and outcomes at nearby and sometimes distant locations. The spatial and temporal context for management decisions should match the scales of ecosystem processes critical to sustainability. Forest ecosystems are constantly changing and such change is often critical to their long-term functioning. This reality is especially important in the drier forest types of the Eastside where fire exclusion has resulted in accumulations of fuel, an abundance of densely stocked young stands, and, consequently, increased risk of wildfire and outbreaks of insects and pathogens.

Complexity and diversity. Management practices for any one species or element must recognize that suitable habitat encompasses all of the other species and ecosystem processes on which that species depends. The area of habitat expected to sustain viable populations of species through time must be sufficiently large to buffer inevitable fluctuations in population size. A landscape or regional approach to distribution of reserves and connections between them is critical.

Uncertainty and surprise. Uncertainty results from complex, often unpredictable interactions among ecosystem elements, limited ecological understanding and poorly developed principles upon which models of ecosystem behavior can be constructed, and poor data quality, sampling bias and analytical errors. Although risks can be reduced, managers cannot eliminate surprises. Adaptive management is critical to dealing with this reality.

Humans as ecosystem components. The effects of human activities on ecosystems–including effects on forest structures and on ecosystem processes—present important management challenges.

Although forest managers have amended management practices to account for changes in markets, social values, uncertainty and risk, and changes in our understanding of the effects of management practices, management must become more flexible in the face of even more rapid changes in human effects on and demands from landscapes and significant changes in our understanding of such factors as context, spatial and temporal scale, and landscape change. Appropriate levels of management must incorporate a well-designed and properly managed system of protected reserves.

> *Appropriate levels of management must incorporate a well-designed and properly managed system of protected reserves.*

> **Recommendation:** A formalized approach for adaptive management should be developed and applied in evaluating the effects of forest management practices on key ecosystem properties and to guide changes in these practices that reflect forest conditions at all spatial scales.

RESEARCH RECOMMENDATIONS

As noted throughout this report, better knowledge is needed for guiding forest management and resolving issues in the Pacific Northwest. An accelerated program of research is needed to fill these gaps; similar gaps exist for other regions of the country as well.

Various institutions and sources of funding play important roles in forest-related research in this region and in the country as a whole. The federal role in funding both in-house and extramural research is obviously very important, but the states, forest industry, and nonprofit organizations also provide research support. The committee believes all of these institutions can take part in supporting and conducting the needed research. In particular, the federal government should substantially strengthen its support for a competitive research grants program that would recognize the broad array of scientific specialties and research organizations that are relevant to current issues involving forest management and conservation.

Specific areas of research in need of increased funding and attention include the following:

- the relationship of natural disturbances to the sustainability of protected and managed Pacific Northwest forests and the extent to which the effects of these disturbances can be simulated by management practices;
- the relative importance of legacies and their role in maintaining forests and regenerating harvested areas, and the extent to which management actions can "create" legacies;
- the role of insects and pathogens in sustaining natural processes in Pacific Northwest forests and factors involved in insect and pathogen outbreaks in the region;
- forest restoration methods and their role in restoring and maintaining forest vitality;
- the impacts of forest-management practices, including timber harvesting, on the production of nonwood forest products, including recreation and special forest products such as wild-grown mushrooms;
- information for making accurate assessments of the impacts of changes in forest practices on regional and local employment and income;

- impacts of changes in forest practices in the Pacific Northwest on biological and nonbiological factors within the region and in other affected regions;
- continued basic research on the biological functioning and interactions of the multitude of life forms present in the Pacific Northwest forests.

The importance of adequate research funding to develop these kinds of information is shown by the duration and intensity of the conflict over the management of Pacific Northwest forests. Much of the conflict has been over differing views of what is happening to the forests and has been based in most cases on poor or insufficient information. The cost of the conflict and the turmoil it brought to the region has been substantial. At the risk of seeming to be unduly sanguine, the committee believes that attention to developing an appropriate information base will help to resolve similar issues in the future.

1
THE PROMISED LAND:
THE LAND OF PROMISE

INTRODUCTION

In his 1938 dust-bowl-era treatise, Richard L. Neuberger described the Pacific Northwest as "the promised land," and certainly the region has appeared to waves of immigrants as a land of great bounty and promise. The pioneers who followed the Oregon Trail were attracted by the wealth of natural resources and the opportunity for a new future. Compared with the exhausted fields and eroded soils of the Midwest, the immigrants of the dust-bowl era found unexploited forest resources in the Pacific Northwest and abundant hydroelectric power that made new technologies and industries practical.

Humans depend on natural and managed ecosystems to provide a variety of commodities and amenities. By exploiting resources, each wave of migrants to the Pacific Northwest significantly altered the landscape and, in doing so, increased its capacity to deliver some goods and services while diminishing its potential to deliver others.

The lures to migrants moving to this region in the past three decades included economic opportunity associated with the urbanization and industrialization of major transportation corridors, such as along the interstate highway from Portland to Seattle. For migrants in recent times, the proximity of natural ecosystems to zones of economic development provided additional attractions, such as clean air and water, abundant recreation, and escape from population centers.

It is doubtful that many migrants would have consciously advocated practices that would have denied their children or grandchildren the opportunities they themselves enjoyed. But little attention was paid to whether the region could indefinitely meet society's demands for its commodities and amenities. Now the optimism of the past has waned, replaced by a pessimism that pits individuals and groups with differing resource wants and needs against one another. The resulting conflicts have exposed the inadequacy of the protocols and institutions needed to resolve disputes across complex boundaries of environments, jurisdictions, ownerships, and cultures.

What changed since Neuberger's optimistic depiction of this region? Three general trends—the loss of the frontier, a changing knowledge base, and changing societal values—are important.

Loss of the Frontier

Nineteenth century romantic writers depicted the frontier as a cornucopia of wealth and resources (Nash 1982). But that frontier is gone. And despite dwindling resources, demand for the goods and services they produce has increased. But satisfying demands for some resources can conflict or compete with the ability to meet the demands for others. The conflict is not usually jobs versus the environment. Typically and increasingly, conflicts are among types of jobs, for example, when logging reduces employment in fishing by altering aquatic habitats, thereby contributing to the decline of salmon stocks, or when environmental effects dilute amenities that attract other industries, jobs, and workers.

Changing Knowledge Base

As the science of ecology has matured, so has our understanding of the consequences of various human activities on the landscape. Although ecologists have been aware of the relationships between spatial scale and number of species for nearly a century, the connection between that relationship and the loss of species as landscapes become fragmented into smaller, disconnected pieces of habitat has only been recognized in the past two decades (Wilson 1992). Such factors as the variability among species in the breadth of habitat requirements and the complexity of the landscape mosaic (e.g., extremes of fragmentation and the character of the disturbed landscape matrix), make quantitative predictions regarding species loss and habitat fragmentation difficult (Rochelle et al. 1999).

Our knowledge base at any given time is provisional and subject to change. In the early 1900s, selective cutting and high-grading were typical forestry practices; later, foresters shifted to clear-cutting. Technological advances and changing markets that permitted economic use of a greater array of species and tree sizes also contributed to this shift. By the 1960s, the first comprehensive studies demonstrating the negative effects of large clear cuts on watershed hydrology and nutrient cycling were completed. The importance of dead woody debris to the functioning of aquatic and terrestrial ecosystems in the Pacific Northwest has been made clear (Harmon et al. 1986; Perry 1998; Aber et al. 2000).

Almost certainly some components of what is considered today to be best practice will be found to be erroneous. However, the time between the acquisition of new knowledge and understanding and their incorporation into natural-resource management often is measured in decades. Public expectations that decision makers and resource managers understand the resources they manage can make it difficult to admit that much management is necessarily experimental and to establish the institutional structures and monitoring systems needed to learn from experiments. Value-laden distinctions between basic and applied research also create barriers to the application of new knowledge.

Changing Social Values

Demand for wood fiber and its derivative products has increased nearly twofold since 1950 and is projected to double again early in the next decade (NRC 1998). At the same time, public interest in sustainability and ecological consequences of some forest-management practices has grown. Population growth and changing societal values have increased interest in and demand for parks, wilderness, and recreation. An ever-increasing variety of forest organisms have become important, including herbs, wild plant foods, and mushrooms. Stocks of anadromous fishes (such as salmon) that spend critical portions of their lives in forest streams have declined, and public concern has been expressed over the quality and delivery of water, landscape appearance, potential loss of indigenous species, and fragmentation of forests. Legislation such as the Clean Water Act and the Endangered Species Act, as well as state regulations, have changed forest management.

This committee identified four general goals or expectations that society has for forested landscapes:

- Sustain viable populations of indigenous species
- Maintain properly functioning ecological processes
- Meet human needs for forest commodities
- Satisfy cultural and aesthetic values

The relative importance of each of these goals has changed considerably over the past several decades. Forest management in the Pacific Northwest on public and private lands has been aimed primarily at meeting human needs for commodities, particularly wood products. Early in this century, the time horizon for management decisions was relatively short, but it has gradually expanded with increased emphasis on the need for sustained yield to encompass meeting needs for wood products over many years. Management goals gradually shifted, especially on public lands, to encompass new needs, including providing wildlife habitat, protecting water quality, and meeting aesthetic concerns. Sustainable populations of indigenous species have become a major goal in the context of the Endangered Species Act, especially on public lands. Today, more attention is being focused on sustaining ecological processes, a goal that was not fully visualized in historic models for sustained yield of wood fiber (SAF 1993).

Forest-management decisions and actions are made at a variety of spatial scales.

Maintaining natural processes and integrity has been recognized by the National Research Council as a key element in "a transition toward sustainability." Accomplishing this transition as a worldwide goal requires integrating global and local perspectives in "place-based understanding of the interactions of the interactions between environment and society" (NRC 1999a). This report focuses on one such place- or region-based understanding and the information needed for it.

Forest-management decisions and actions are made at a variety of spatial scales. Typically, forest-management decisions focus at the stand or site scale (measured in acres or hectares) and on such issues as the size of a cut or specific harvest practices like clear cutting. Increasingly, management is concerned with landscape and watershed issues such as the relationship among management patches and the connections between them, or the accumulated consequences of multiple activities within a watershed. And many critical decisions and actions, such as the management of animals that migrate over large distances or management decisions that influence patterns of human development or extensive accumulation of flammable fuels, are taken or have significant consequences at the scale of regions (for example, entire states or subregions, such as the Eastside or Westside).

A recurring theme in Pacific Northwest forest management is that reasonable goals set or actions taken at one spatial scale can have unfavorable consequences at another scale. Best-management practices can be applied at the scale of individual stands, but if attention is not paid to the spatial arrangement among stands, such practices might have negative effects on the hydrological flows in watersheds or on populations of wide-ranging wildlife species. Such spatial mismatches have analogues in the temporal dimension; e.g., reasonable management decisions from the standpoint of a fiscal year or electoral cycle can diminish long-term capacity and sustainability. If there is one overarching lesson from the current management dilemmas, it is that mechanisms and institutions are needed to reconcile management goals and actions over scales of space and time.

More people want more things from forests, and societal priorities with regard to the variety of goods and services provided by forest ecosystems have clearly shifted. A few people were concerned about the loss of species from forests 50 years ago, but only recently has worldwide loss of biodiversity has become a mainstream public concern. The ancient forests of the Pacific Northwest have become a major symbol of that concern. People across the country who might never visit an old-growth, Douglas-fir forest have lobbied policy makers and provided financial support to various nongovernmental organizations, thereby becoming important stakeholders in decisions affecting the fate of those forests.

So far, existing institutions and attempts at conflict resolution have failed to achieve a common societal vision for the Pacific Northwest.

People's wants and needs often exceed the ability of ecosystems to meet them, and this remains the most significant challenge to achieving a workable vision. Borders and patterns of human ownership and jurisdiction do not generally correspond with the spatial scales and boundaries that define integrated hydrologic systems or the behavior of wildlife and wildfire. Society's desires for particular outcomes at landscape, regional, and even national and international scales often conflict with individual wishes to achieve certain benefits at local scales or established rights to the use of personal property. Time scales of fiscal years and 10-year management plans often drive management decisions, whereas ecosystem processes that sustain the supply of goods and services operate over many decades and centuries. Consequently, the results of actions and practices often do not become evident until long after they are applied.

The future of the Pacific Northwest is uncertain given the numerous debates and proposed adjustments in forest management, but the outlook is positive. Although concerns about unsustainable patterns of land use and forest fragmentation are real and legitimate, more significant expanses of relatively undisturbed forest and wilderness remain than in other regions of the United States. Past forest management has created threats to the well being of some timberlands, but most of the Pacific Northwest second-growth forest remains productive. Furthermore, management practices continue to improve, and new technologies and protocols are being implemented that promise to diminish the adverse effects of extractive practices. Important elements of the rich biodiversity of the Pacific Northwest, including some of considerable economic value, are indeed threatened. However, opportunities exist for recovery of populations and restoration of habitat.

> *The future of the Pacific Northwest is uncertain given the numerous debates and proposed adjustments in forest management, but the outlook is positive.*

Some timber-dependent communities have suffered economically as the flow of old-growth timber has been curtailed, but the Pacific Northwest economy is, on the whole, vibrant. Most small rural communities have made transitions to a diversified economic base and are benefitting from the overall well being of the region, although important segments of the region's rural population endure hardship and substandard living conditions.

The vast expanses of federal land in the Pacific Northwest provide opportunities for landscape management and old-growth forest preservation that might not be available in other regions of the country, but those opportunities can be realized only if management practices on nonfederal lands are also taken into account. An integrated regionwide approach to forest management that recognizes the opportunities for resource use across the spectrum of ownerships should provide the potential for true win-win outcomes. However, few institutions or structures exist to plan for or effect decisions on those scales.

Forest products enter common economic markets, whether the products are extracted from public or private lands or whether they come from the Pacific Northwest, other regions of the United States, or other countries. Decisions that influence supply from one ownership or region necessarily influence management decisions of others. If sustainable provision of the functions, goods, and services provided by our forested landscapes is to be achieved, consideration should be given to coordination across local, regional, and global scales. Given the growth of human populations, as well as increased global per capita consumption, there is little doubt that worldwide demand for wood and wood fiber will continue to increase. Through its actions and policies, the United States has an opportunity to encourage and set an example for sustainable forest ecosystem management for the rest of the world.

It is still rational to view the Pacific Northwest as a land of promise, but the region's ecosystems can deliver the goods, services, and amenities on which humans depend only if people fulfill their collective responsibility for wise stewardship. This requires seeing forest landscapes and resources more as a trust held for our children than as expendable resources inherited from our ancestors.

THE STUDY AREA

To address the conflicts in and problems symbolized by the Pacific Northwest, Congress asked the National Research Council to review forest-management practices, examine old-growth forest issues, and identify the current status of knowledge. The NRC convened the

Committee on Environmental Issues in Pacific Northwest Forest Management; this report is the culmination of that group's deliberations. The committee defined the Pacific Northwest as the states of Oregon, Washington, and Idaho, the northern part of California, and Montana west of the crest of the Rocky Mountains (Figure 1-1). This region includes the major forested ecoregions of the three Northwest states and their extensions into northern California and Montana. Boundaries were delineated as follows: the entire states of Oregon and Washington, the U.S. portion of the Columbia River drainage basin (which contains salmon habitat), the Klamath and northern coastal regions of northern California (because they are an extension of the ecoregion of southern Oregon and contain part of the northern spotted owl habitat and salmon habitat). Because not all data sources use the same geographic base, the exact borders of the region used for particular analyses and comparisons vary throughout the report. For example, the demographic analyses may use different counties than those used in reports of timber products. In most cases, the variations are minor and did not influence overall conclusions.

For some discussions, the committee further divided the Pacific Northwest into three geographic subregions that reflect the major ecoregions in the Pacific Northwest: Eastside, Westside, and Northern Rocky Mountains. The division into subregions follows standard practice in the forestry literature and in the way in which the U.S. Forest Service collects and publishes forest resources data. The Westside is the high precipitation area west of the Cascade crest from the Canadian Border south to northern California. The much drier Eastside extends from the Cascade crest east through eastern Washington and Oregon and central Idaho. Forest conditions, including species composition, between the Westside and Eastside differ substantially. The northern Rocky Mountain area of northern Idaho and western Montana has higher precipitation than the Eastside, has a more continental climate, and has different species composition of the forests from that of the other regions.

OTHER MAJOR STUDIES AND REPORTS

Most of the many studies and reports that have addressed various aspects of Pacific Northwest forest management dealt with

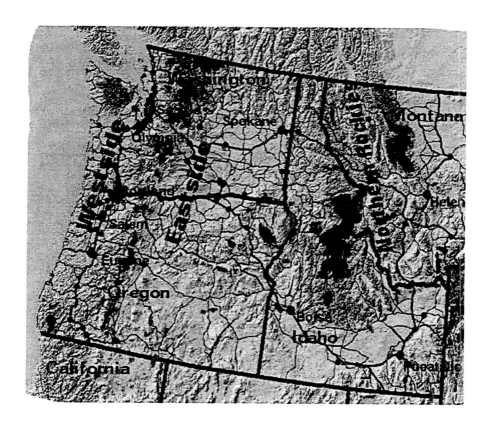

FIGURE 1-1. The Pacific Northwest.

forest-management issues on the Westside. A series of reports in the 1960s and 1970s focused on the region's timber supply situation USFS 1963, 1969, 1976; Beuter et al. 1976). A major concern was whether federal timber harvests could be increased during the last decade of this century and the first couple of decades of the next while private forests that had been logged grew back to regain their place as the main source of the region's timber.

In the 1980s, attention turned to habitat protection, notably protection of the northern spotted owl (*Strix occidentalis caurina*), a threatened species that depends primarily on old-growth forests. Then attention expanded to protection of other species, including the marbled murrelet (*Brachyramphus marmoratus marmoratus*), a seabird that nests in old-growth forests, and several salmon stocks. Concern with the loss of old-growth habitat prompted a new series of studies.

The federal Interagency Scientific Committee (ISC) was the first of several important and innovative scientific efforts to bring economic and ecological stability to the Pacific Northwest. ISC members, appointed by the chief of the Forest Service, proposed in 1990 that large reserve blocks on federal forests capable of supporting 20 or more pairs of spotted owls replace the previous strategy of protecting spotted owl habitat areas around each nesting pair of owls. The ISC recommended habitat conservation areas (HCAs) for spotted owls and that timber harvesting be prohibited in HCAs. In response to court decisions that stood in the way of effecting this proposal, a congressional committee appointed the Scientific Panel on Late-Successional Forest Ecosystems to examine a range of alternatives (Johnson et al. 1991). It reported that adequate protection of species partly dependent on old-growth forests, especially fish such as salmon, would require a greater degree of protection than was proposed in the Forest Service plan.

With the issue of habitat protection still unresolved, the Forest Service chartered another study by the Scientific Analysis Team (SAT 1993). That group recommended that habitat conservation areas be combined with riparian protection zones to protect old-growth dependent species and aquatic species.

After an April 1993 forest conference in Portland, Oregon, President Clinton created the Forest Ecosystem Management Assessment Team (FEMAT) to develop and analyze several options, all of which included extensive reserve systems. Those options were to be scientifically credible and legal, with minimal negative social impacts. The president announced his choice of "Option 9" on July 1, 1993, and this was followed by an environmental impact statement with a revised Option 9, known as the Northwest Forest Plan (NFP), which was adopted in February 1994. The NFP required protection of late-successional old-growth reserves and riparian reserves on federal land. It allowed some silvicultural activities in the reserves, extended timber rotation ages to 180 years on nonreserved federal land in California, required buffer zones to protect the marbled murrelet, and provided for increasing the amount of coarse woody debris and green trees left following logging on nonreserved areas. One effect of the NFP reserves is to reduce the probable timber harvest on federal "owl forests" to about a quarter of what it was during the 1980s. The NFP also includes a controversial "survey and manage" provision that requires extensive surveys of plant and animal species potentially affected by timber harvest. The NFP allows silivicultural activity in stands less than 80 years of age that are in late-successional reserves, if the activity is designed to accelerate the development of late-successional forest conditions. A substantial portion of the timber harvest since adoption of the plan has come from the late-successional reserves. The NFP was accepted by the responsible federal district court in December 1994 as meeting legal requirements.

Even as the process leading to the NFP, which dealt primarily with Westside forests, was moving along, attention was also being directed at somewhat parallel concerns on the Eastside. In response to a congressional request, the Forest Service produced a report, *Eastside Forest Ecosystem Health Assessment* (USFS 1993a). A more recent report addresses issues of fisheries management for anadromous fish throughout the Pacific Northwest, Eastside, and Westside (PACFISH 1994). Neither report has led to actions similar to the adoption of NFP, but the concerns broadly parallel those addressed by FEMAT.

Since FEMAT and the Eastside Assessment, much attention has been given to the health of Pacific Northwest forests, particularly with respect to the impacts of past fire suppression and the possible benefits of timber harvest as a surrogate for natural disturbance. Oliver et al. (1997) argued that forest harvesting should be increased in many regions, including the Pacific Northwest, in order to reduce fuel loads and the danger of catastrophic fires. Others (SNEP 1996a, b; Aber et al. 2000) argued that fuel accumulations owing to fire suppression have been overstated in many areas and warned that harvest activities are often a poor substitute for natural disturbances such as fire and windthrow.

In 1998, the National Research Council Committee on Prospects and Opportunities for Sustainable Management of America's Nonfederal Forests (NRC 1998) outlined strategies to improve the health of forest ecosystems on private land and improve the incentives to private land owners for sustainable management. The Committee of Scientists Report (1999), written by a committee of non-Forest Service scientists appointed by the Secretary of Agriculture to review the current regulations for land-use planning on the national forests, emphasized the importance of ecological sustainability and public participation in the management of national forests.

THIS REPORT

This report describes the Pacific Northwest and its forests (Chapter 2); presents information on the status of the regon's biological and hydrological resources (Chapters 3 and 4); examines the various definitions of old-growth forests (Chapter 5); reviews the way in which changes in the use of forest products from the region affect supplies from other regions of the country and the world (Chapter 6); presents information on the effects of forest management on human communities in the region (Chapter 7); and reviews forest management practices in the region, their effects, and alternative management approaches (Chapter 8). The final chapter summarizes the committee's conclusions and presents its recommendations (Chapter 9).

Clear goals are essential to any effort to rationalize forest management in the Pacific Northwest. The four goals formulated by the committee are at the heart of the issues in forest management in this region (see page 18 and Chapter 8). Those goals provide the general framework and helped to clarify the kinds of issues that are involved in Pacific Northwest forest management. But conflicts arise among them because they cannot all be maximized or optimized simultaneously. The citizens of the region and the country must decide what weights should be assigned to each of these goals.

2
THE REGION AND ITS FORESTS

INTRODUCTION

This chapter presents a brief history of the development of the Pacific Northwest region in relation to its forests and their uses, a current demographic and economic profile of the region in relation to its forests, and a short description of the forests themselves.

The region's forests are complex and vary depending on precipitation, soils, elevation, disturbance patterns (e.g., fire, wind, disease, and insect infestations), management, and use. The forests are traversed by transportation and communication networks and are separated in places by broad expanses of cropland, rangeland, and urban areas. They consist of a mix of ownerships, public and private, that have been subjected to quite different management regimes (Table 2-1).

A BRIEF HISTORY

The current status of the forests, the forest industry, or the society of the Pacific Northwest can be understood only in the context of its history. Humans have probably been altering the ecology of forests in the Pacific Northwest for more than a millennium (USFS 1993a). For example, on the east side of the Cascade Range, American Indians burned the hillsides to improve the production of black mountain huckleberries, blueberries, and grouseberries. Those fires created openings that attracted deer and elk. The landscape was changed further by the 1800s, when native peoples in the interior Northwest acquired large numbers

TABLE 2-1. Area of Forest Land in the Pacific Northwest (Thousands of Acres)[a]

	Timberland[b]						Reserved timberland[b]	Other forest	Total forest	Total land
	National forest	Other federal	Indian	State and local	Forest industry	Other private				
California										
North interior[c]	3,190	126[e]	–[e]	–[e]	1,757	580	554	3,606	9,812	13,992
North coast[d]	619	263[e]	–[e]	–[e]	1,301	1,430	200	1,002	4,815	6,209
Idaho	9,705	605	38	1,087	1,198	1,901	3,051	4,234	21,818	52,961
Montana[f]	4,778	60	344	443	1,361	887	1,200	–	9,073	13,627
Oregon	10,152	2,310	315	929	5,114	3,265	1,777	4,196	28,057	61,442
Washington	4,859	167	1,376	2,250	4,588	3,609	2,765	2,244	21,856	42,612
Total	33,303	3,531	2,073	4,709	15,319	11,672	9,547	15,282	95,431	190,843

[a]The region is defined along political boundaries to include all of Idaho, Oregon, and Washington, counties on Montana west of the Continental Divide, and northern counties in California. Area statistics are from selected USFS reports: Waddell et al. 1989 (Id., Ore., and Wash.), Coulclasure et al. 1986 and Lloyd et al. 1986 (Calif.), Jackson and Jackson 1987 (Mont.).
[b]Timberland is forest land capable of continuously producing ≥ 20 ft^3/ac/yr of industrial wood and not withdrawn from timer buse by statute, ordinance, or administrative order. Reserved timberland is otherwise productive forest that has been withdrawn.
[c]Lassen Modoc, Shasta, Siskiyou, and Trinity counties.
[d]Del Norte, Humboldt, Mendocino, and Sonoma counties.
[e]Indian forest land and state- and local-government forest land is included under other federal.
[f]Montana statistics include forest land from the Bitterroot, Flathead, Kootenia, and Lolo national forests and all other forest land in Flathead, Granite, Lake, Lincoln, Mineral, Missoula, Ravalli, and Sanders counties.

of horses. Although the Indians may have altered the landscape in some areas, the consequences were probably minor in comparison with natural disturbances such as fire, storms, and drought.

The first European pioneers into the Pacific Northwest followed the river valleys: today, most major cities are along those same transportation routes. With the arrival of white settlers in the early 1800s, changes in the region accelerated. Forests were cut to clear land for houses, farms, and towns. Timber, minerals, hydroelectric power, fish, range livestock, tourism, and agriculture all played important roles in the region's development.

The federal government also played an important role in the development of the region's resources. To encourage settlement of the Pacific Northwest, resources on public lands and the land itself often were given away or sold at low prices, as was true throughout the West. In the case of hydroelectric power and irrigated agriculture, the federal taxpayer bore much of the cost of development.

The basic patterns of land ownership and subsequent use were established in the mid-nineteenth century. Federal public-domain lands acquired in the Oregon Compromise of 1846 were sold or granted to encourage settlement and development of the region. Still other lands were reserved from disposition to form national parks and national forests, and Indian reservations were established in treaties between the United States and sovereign Indian nations.

Large areas were transferred to the states or private ownership (Gates 1968). Statehood grants to support schools—two sections[1] per township of 36 sections—have for the most part been retained by the states and are managed in trust for the schools. But other grants to the states for railroads, wagon roads, and other internal improvements, as well as swamplands, were generally sold to settlers and others. The total area of grants to state governments was substantial: 7.9% of Idaho, 11.4% of Oregon, and 7.1% of Washington. In addition, substantial areas went directly into private ownership under various laws, including the homestead acts, the 1878 Timber and Stone Act, and the 1850 Oregon Donation Act.

Land grants directly to railroads have had a major effect on forest land-use patterns (Gates 1968). The Northern Pacific Railroad received

---

[1] One section is 1 square mile or 640 acres.

a grant of 40 alternate sections (25,600 acres) for each mile of railroad across Montana, Idaho, and Washington in a band as much as 60 miles wide on both sides of the right-of-way. Some of that land in forested areas was sold to timber companies, and some is still held by timber firms descended directly from the original companies. Those sales and other grants to railroads in the region created large regions of checkerboard ownerships, where alternate sections of private lands are intermingled with federal forest lands.

In Oregon, 2,891,000 acres of a grant to the Oregon and California Railroad was revested to the federal government in 1916 when the company failed to meet the terms of the grant (Gates 1968). About 93,000 acres of a grant to the Coos Bay Wagon Road were also returned to the federal government in 1919 (Richardson 1980). Those two areas, together with a small amount of forest land that was never appropriated for private use and was not included in areas reserved for national forests, make up the highly productive Oregon forests managed by the Bureau of Land Management (BLM).

Today, about half the land in the Pacific Northwest is in public ownership, and most of that land is managed by the U.S. Forest Service (USFS) and BLM. State governments and a variety of other federal agencies also manage public lands in the region. American Indian tribes and the Bureau of Indian Affairs (BIA), manage large tracts of land on reservations scattered throughout the region (Table 2-1). Private owners have holdings that range in size from small woodlots to millions of acres of industrial forests owned by large corporations.

The forest resources in the Pacific Northwest have provided income and basic materials for a growing population. The U.S. National Resources Committee called forests "the chief means of payment for the products of other areas" (USNRC 1938), and forests were the most important factor in the development of the Pacific Northwest through the mid-1930s. Although the main economic commodities from forests are timber and timber products, Pacific Northwest forests provide a variety of other goods and amenities, such as wildlife, recreation, water, wilderness, edible berries and mushrooms, ornamental plant materials, and medicines.

By the early twentieth century, the Pacific Northwest had become the country's leading lumber-producing region (USNRC 1938); by 1930, forest industries provided 41% of the value of the region's net exports.

The dominance of forest products was even greater in Oregon and Washington, where they accounted for 64% and 54% of the value of the net exports, respectively.

Increasing affluence after World War II led to a large increase in the demand for wood. Forests began to be viewed as renewable resources capable of supplying a more or less continuous flow of timber, thereby eliminating the boom-and-bust cycles that had characterized the logging industry before the war. In the 1960s, forest management became more intensive, with many forest owners looking for ways to increase yields and shorten rotations (the intervals between harvests). Management strategies such as artificial regeneration — mainly tree planting — became common practice; this was followed in the 1970s by other strategies, including tree breeding for genetic improvement, fertilization, thinning, and pruning. New logging practices were developed to lower costs of logging and road building.

The place of federal forests in supplying timber for the wood-products industry in the Pacific Northwest has varied. Private forests provided most of the timber until the 1950s. During the 1930s, the timber industry actually lobbied to hold down federal timber harvests in western Oregon because of fears about their possible effect on private timber prices (Robbins 1985; Richardson 1980; Steen 1976). Federal forests provided an increasing share of the region's timber harvests after World War II as lumber and plywood production grew to meet the demand for residential construction. From 1952 to 1976, total softwood timber harvests in the region increased by 37%, but harvests from the national forests went up by 87% (USFS 1982).

> Today, about half the land in the Pacific Northwest is in public ownership, and most of that land is managed by the...USFS and BLM.

Prompted in part by concern over the rapid increase in national forest timber harvests, Congress passed the Multiple-Use, Sustained-Yield Act in 1960, which, among other things, carried the tacit promise that timber harvest levels would be maintained into the future (Steen 1976; Hagenstein 1992). Historically, most people in the timber industry thought of sustained yield only in terms of wood; many still do (Hammond 1991; USFS 1991). But the multiple-use mandate extended to resources and uses other than just wood, including wilderness, outdoor recreation, wildlife and fish, and grazing (NRC 1990, 1998).

Throughout the post-World War II era, future timber supplies were a concern in the Pacific Northwest, especially in view of the increasing share coming from the national forests. The research branch of the USFS examined the implications of the expected change from harvesting mainly old-growth timber in the Douglas-fir subregion to relying on second-growth timber. It appeared certain that public forests would have to provide an increasing share of the region's timber harvest if the total harvest were to be maintained (USFS 1963). Harvests from national forests in the region, however, were constant from 1960 until about 1990. Some strenuous efforts were made to increase these harvests substantially, but the USFS generally resisted those efforts and held firm to its nondeclining, even-flow policy.

But even without increases in timber harvests, it was becoming increasingly difficult to maintain a constant yield of timber from the federal forests. Shortly after the sustained-yield policy was established in law, Congress began chipping away at the potential contribution of the national forests to the timber supply, first with the Wilderness Act of 1964, and then with a series of other designations that effectively withdrew parts of the national forests from timber production (Leshy 1992).

Despite such restrictions on the timber supply, the region's total output of wood products has increased since 1950. Lumber production reached its highest level in 1946 and stayed close to that level through the late 1980s. Plywood production increased rapidly after World War II, although it has fallen from its peak in the 1970s. Log exports were a minor drain on the region's forests until about 1970 but accounted for 10-15% of the region's timber harvests from then to the late 1980s. Although total wood-product output has increased since 1950, timber harvests have not increased as rapidly because of steady decreases in the waste of timber used in processing (Adams et al. 1988).

Despite relatively heavy harvests over the past several decades, substantial timber volumes remain on lands in all ownership categories (Waddell et al. 1989). In 1992, the average volume of sawtimber (trees suitable for making lumber or plywood) in forests available for timber harvesting in Oregon and Washington was more than 18,000 board feet per acre (Table 2-2).

The Region and Its Forests

TABLE 2-2. Average Volume in Board Feet of Sawtimber Per Acre by Ownership Category, 1992

	Boardfeet		
	Idaho	Oregon	Washington
National forest	10,475	20,453	24,198
Other public	9,615	28,941	22,086
Forest industry	8,462	10,736	14,875
Other private	6,281	7,794	8,809

Source: USFS 1993a.

This compared with about 3,900 board feet per acre in the southern United States. Of the Pacific Northwest's total timber not formally reserved from harvest, private lands had 29%, national forests had 52%, and other public lands had 19%.

DEMOGRAPHICS AND THE ECONOMY

Population Growth

Since 1970, the Pacific Northwest has undergone numerous dramatic changes, including economic and cultural changes and significant increases in human population. Today, about 4% of the U.S. population lives in the Pacific Northwest.

A wave of immigration began shortly after World War II when large numbers of people moved to the Pacific Northwest, mostly to its two largest metropolitan areas, Seattle and Portland. A second wave of immigrants, this time largely retirees, began in the mid-1970s. Communities with distinctive recreational and aesthetic assets, such as coastal towns like Brookings and Brandon in Oregon, and inland communities like Bend, Oregon, and Coeur D'Alene, Idaho, grew rapidly. By the 1980s, the Pacific Northwest was adding to its population at a rate of 1.9% per year, twice the U.S. national average of 0.99% (U.S. Bureau of the Census 1990).

The metropolitan areas and cities on the Westside account for most of the population of the Pacific Northwest and dominate the region's economy (see Figure 2-1 and Figure 2-2). Westside residents earn most of the region's personal income: 69.4%. The economic and demographic

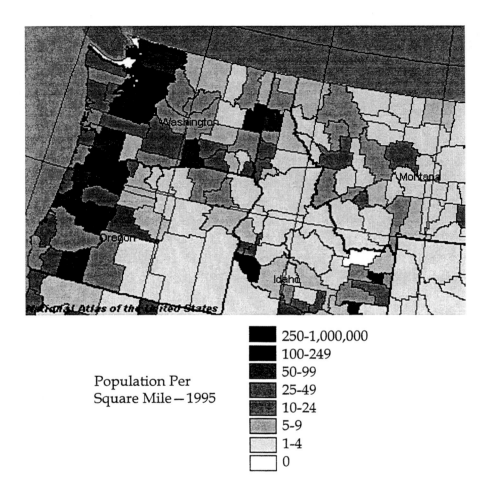

FIGURE 2-1. Population in the Pacific Northwest.

The Region and Its Forests

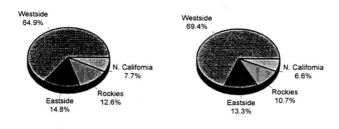

a) 1991 Population   b) Total personal income

FIGURE 2-2. a) 1991 population and b) total personal income, by subregions. Data from the U.S. Bureau of the Census. Areas included are Oregon, Washington, and Idaho and counties in California and counties in Montana.

dominance of the Westside has concentrated political strength and recreation demand in that area. Across the Pacific Northwest, increasing urbanization has coincided with a shift in public values—greater absolute demand for nonwood forest products (such as environmental amenities, wildlife, parks, recreation, and wilderness) and greater demand for nonwood forest products relative to wood products and other extractive uses of forest lands.

Economic Implications of Population Growth

Rapid population growth has been accompanied by two significant changes in the regional economy. First, the aging of the population and growing numbers of retirement-age immigrants has increased the importance of nonlabor income (e.g., interest and dividends). The second major change in the regional economy is its diversification. The proportion of total personal income generated by the wood products industry declined from 6% in 1971 to 3% in 1992. At the same time, other manufacturing industries increased their shares of total personal income, in sharp contrast to the national trend toward decreased importance for manufacturing.

Adjusted for inflation, personal income in the Pacific Northwest has risen at an annual rate of 3.2% since 1971. This reflects very strong growth in nonlabor income sources (4.5% annually) and more moderate growth in labor-derived income (2.7%).

> *The regional economy of the Pacific Northwest is remarkably similar to the that of the nation.*

The regional economy of the Pacific Northwest is remarkably similar to the that of the nation. Manufacturing industries account for nearly the same total share of the gross product (the value of all goods and services produced) for the Pacific Northwest as for the nation (approximately 19% in 1990). Transportation, communications, public utilities, and trade in the region produced 33.6% of the gross regional product; it was 32.4% of the gross national product. The Pacific Northwest is less reliant on service industries than the nation as a whole — 15.7% of the gross regional product compared with 17.9% of the gross national product.

Even with changes in the regional economy, the forest products industry is an important export base. The region's share — more than $16 billion — of the nation's total annual value of shipments by the lumber and wood products industry is about 6 times the region's share based on indicators of demand for wood products, such as population, personal income, and value of construction contracts (U.S. Bureau of the Census 1978; U.S. Department of Commerce 1993). This is a strong indication of the importance of wood products as a source of income for the region.

THE REGION'S FORESTS

Westside Forests

The Cascade Mountains are largely responsible for the temperate climate and generally moist conditions on the Westside. Much of this area receives more than 254 cm of precipitation per year. Precipitation is strongly seasonal, with 75% occurring between November and March, with only small amounts between June and October. Because hardwoods are less drought tolerant, the pattern of winter rain and summer drought favors growth of conifers over hardwoods (Waring and Franklin 1979). The moist climate west of the Cascade crest also makes fires infrequent and fire-return intervals relatively long and highly variable.

Coastal areas have dense temperate rainforests dominated by Sitka spruce (*Picea sitchensis*), western hemlock (*Tsuga heterophylla*), western red cedar (*Thuja plicata*), and grand fir (*Abies grandis*) and Pacific silver fir (*Abies amabillis*) in the north, grading into redwood (*Sequoia sempervirens*) forests in southern Oregon and northern California. Interior areas are predominantly western hemlock, Douglas-fir (*Pseudotsuga menziesii*), and western red cedar up to midelevations in the Cascades. Red alder (*Alnus rubra*) is common at recently disturbed sites, and western red cedar is characteristic of particularly wet areas. True firs (*Abies*) and mountain hemlock (*Tsuga mertensiana*) dominate at higher elevations (Franklin and Dyrness 1973).

The Klamath and Siskyou mountains of southwestern Oregon and northern California have very old and diverse exposed strata and an east-west orientation of ridges that funnel weather patterns inland. Those factors and the intersection of several vegetative zones make this possibly the most biologically diverse section of the United States (Whittaker 1960). Gradients in species composition with elevation and aspect are particularly pronounced here, with mesic assemblages at upper elevations and on northern slopes and arid assemblages at lower elevations and on southern slopes (Whittaker 1961).

The amount of Douglas-fir relative to western hemlock in an area depends on disturbance, especially fire, and on the amount of moisture. Where fires occur at intervals of 100 to 400 years, Douglas-fir is dominant. Less frequent fires (intervals of more than 600 years) favor hemlock forests (Agee 1993; Huff 1984). In moist areas, hemlock may also predominate.

The midelevation (1,000 to 2,500 m) forests on the Westside are dominated by Douglas-fir, western hemlock, Pacific silver fir, noble fir (*Abies procera*), and red fir (*Abies magnifica*). At lower elevations, Douglas-fir and western hemlock also can be present, and at higher elevations, mountain hemlock and yellow cedar (*Chamaecyparis nootkatensis*) are common. Above 2,500 meters, the subalpine forests have open canopies of mountain hemlock (associated with lodgepole pine (*Pinus contorta*), whitebark pine (*Pinus albicaulis*), and other fir species); in more continental climes and on the east slopes, subalpine fir (*Abies lasiocarpa*) (associated with Engelmann spruce (*Picea engelmannii*)), lodgepole pine, whitebark pine, and incense cedar are common.

The Douglas-fir/hardwood and Douglas-fir forests are characteristic of areas in northwestern California and southwestern Oregon, with the latter forest type extending along the east slope of the Cascades (Agee 1991). The upper canopy is predominately Douglas-fir; tanoak (*Lithocarpus densiflorus*), canyon live oak (*Quercus chrysolepis*), golden chinquapin(*Castanea pumila*), and Pacific madrone (*Arbutus menziesii*) make up the understory.

Eastside Forests

The landscape and its vegetation are highly varied on the eastside of the Cascade range. The highest mountain elevations (more than 2,500 m) have harsh temperatures, and high winds limit the number of species. Below 2,500 m, frequent drought and fuel accumulation favor short fire-return intervals (including low-intensity surface fires) (Agee 1993). The Eastside receives an average of 25-51 cm of rainfall annually and as much as 81 cm at higher elevations, such as in the foothills of the Rocky Mountains north of the Snake River. Some of the subalpine forests of the Westside also extend east of the Cascade crest.

Ponderosa pine (*Pinus ponderosa*) forests are characteristic of the lower elevations of the eastern Cascade Mountains and intermountain ranges and extend to lower elevation grasslands and shrublands. The forests often have open canopies with a heterogeneous understory of grasses and shrubs. Lodgepole pine is typically found at higher, moist elevations. The importance of fire in this zone is well established (Franklin and Dyrness 1973). Suppression of fire over the past century in these forests has favored growth of the shrubby understory species and invasion of shade tolerant firs and Douglas-fir.

Well east of the Cascades, western white pine (*Pinus monticola*) forest extends from southern Canada to the Locksa Divide in central Idaho, east on better soils into northwestern Montana, and west to a boundary with ponderosa pine in western Idaho and eastern Washington (Larsen 1930). Much of the original white pine forest has been logged or killed by white pine blister rust (*Cronartium ribicola*), an introduced pathogen. The white pine forest is richer in plant species than any other forest type in the region east of the Cascades (Larsen 1930). Various other conifer species commonly grow with white pine, particularly shade-tolerant species, such as western red cedar, western hemlock, grand fir, and Engelmann spruce.

The grand fir zone extends along the eastern Cascade Mountains and the Blue Mountains into somewhat drier areas. Douglas-fir and ponderosa pine are common with grand fir in this zone; Douglas-fir tends to be more prevalent in Idaho. Ponderosa pine, western larch(*Larix occidentalis*), lodgepole pine, western red cedar and western hemlock are also typical of this zone (Frenkel 1993). A white fir (*Abies concolor*) zone, with ponderosa pine and Douglas-fir, extends from the Sierra Nevada north into the eastern Cascades of Oregon. It grades into forests dominated by red fir at higher elevations.

Northern Rocky Mountains

The Northern Rocky Mountain area is an extremely complex ecosystem with regard to climate, substrate, and biota. From the west, elevations increase eastward to the Bitterroot Divide, then decrease eastward into Montana, finally increasing again toward the Continental Divide of the Rocky Mountains. Weather systems usually originate over the Pacific Ocean southwest of the region, and there are rain shadows (areas on the leeward side of mountains) between the Cascade range and the Bitterroot range and east of the Bitterroot range.

Annual precipitation decreases progressively toward the southern portions of the region, as well as from higher to lower elevations (Franklin and Dyrness 1973). About half of the annual precipitation is snow, and precipitation is light in summer. There is a north-south gradient in climate: northern areas are more maritime and southern areas are more continental in climate. Steep mountainous slopes create a dramatic solar-energy flux with large diurnal temperature variations. Cold-air masses collect in drainages and valley bottoms, often creating an inversion of the normal vegetation gradients.

The forests of this ecoregion are influenced by catastrophic, stand-destroying wildfires. This area includes the western red cedar zone and extensions of the grand fir, Douglas-fir, and ponderosa pine zones. Western hemlock, grand fir, and western red cedar communities dominate the more mesic northern portions of the region, and lodgepole pine, Douglas-fir, and ponderosa pine forests are common farther south. High-elevation forests include subalpine fir and Engelmann spruce communities, with lodgepole and whitebark pine communities also common. Quaking aspen is widely distributed across the region but is most common in the drier forests to the south. The zone in northern Idaho and western Montana is predominantly western red cedar, western hemlock, and western white pine. Grand fir and western larch are found in drier sites.

REGIONAL OWNERSHIP PATTERNS

Forest landownership is important because of the degree of control over management that can be exercised by the owner, the differences among owners in their objectives, and the regulations to which the owners are subject. The region's forest ownership pattern was established early as population centers and transportation corridors were developed, public-domain lands were granted to the states and railroads, Indian reservations were created, and the major federal land systems (national forests and parks) were reserved from the public domain. Those patterns are unlikely to be changed greatly in the foreseeable future, although development is sure to spread. Five broad ownership categories are especially relevant: federal, other public (mainly state), Indian, forest industry, and nonindustrial private.

Of the federal forest lands, the national forests are most extensive, but the BLM holdings and the national parks are also important for issues addressed in this report. Forests on Indian reservations, which are private but sometimes viewed as quasi public, are important in the region because they are fairly extensive. State-owned forests are of moderate importance, except in Washington, where they are extensive and have been consolidated to a substantial extent. The two big categories of private forest land, forest-industry and nonindustrial, are important not only because they are extensive and are managed to meet various private objectives, but also because of their juxtaposition with federal forest lands.

Congress has used its plenary authority over the federal lands to designate national parks, wilderness areas, wild and scenic rivers, national recreation areas, and other kinds of jurisdictions and uses. Many of these were created from forest land in national forests. Although Congress was specific in setting uses of these designated areas, it has left the allocation of land uses on the remaining national forest and the BLM holdings to the various land-management agencies.

> *Five broad ownership categories are especially relevant: federal, other public (mainly state), Indian, forest industry, and nonindustrial private.*

USFS and the BLM allocate land uses under broad multiple-use statutory guidelines that give the agencies wide discretion in determining specific uses or mixtures of uses. The planning process identifies streamside and scenic influence zones, special wildlife habitat, and intensive recreational-use areas, as well as timber management and harvest areas. Public involvement in this planning is intended to ensure land allocations and other management decisions reflect broad public interests. Plans are reviewed every 10 to 15 years to allow adjustment for changing conditions.

The various reserves[2] and riparian zones that are emerging as part of the Northwest Forest Plan will override allocations in existing plans, but not the congressional designations of national wilderness areas, parks, and recreational areas.

Most of the state-owned forest land in the region was granted to support schools as part of statehood acts. Acting as trustees for the schools, the states generally manage the lands to maximize income from timber production. As other demands on the lands have mounted, the states have responded with management plans that recognize recreational, environmental protection, and other uses, even where those might compete with timber production.

---

[2] For the purposes of this report, a reserve is any public land on which commercial timber harvest is prohibited, such as wilderness areas, national parks, and the reserves created in the NFP.

Under its trust responsibility for managing forests on Indian reservations, the federal government (through the Bureau of Indian Affairs) works with some tribal governments in planning the use and management of Indian forests. BIA tends to see its responsibility as maximizing income for the tribe, which usually translates to an emphasis on timber production wherever possible. The tribes, on the other hand, are often more interested in cultural values (IFMAT 1993). Many tribes with large areas of forest plan and conduct their own forest management programs and also work with federal, state, and local governments to manage lands affected by the Endangered Species Act and other federal acts.

On private forest land, each owner decides on the uses and management. Forest-industry firms tend to take a fairly long-range view in selecting management goals and programs. The objectives of nonindustrial private forest owners are more variable. Ownership tends to be less stable and the time perspective shorter than for industry firms (NRC 1998). And statewide land-use controls, such as those of the Oregon Land Conservation and Development Commission, can limit an owner's ability to change from forest to nonforest uses, although such controls typically do not require the owner to devote the land to particular forest uses (Diamond and Noonan 1996).

The mix of ownerships and purposes for which forests are managed and used in the Pacific Northwest amplifies the effects of the natural patchiness of forest landscapes. The intermingling of forests and ownerships presents a complicated situation for management of Pacific Northwest forests.

SUMMARY

Human activities have changed the face of the Pacific Northwest considerably. Today's landscape is a mix of public and private lands put to diverse uses; the Pacific Northwest is well known for its checkerboard ownership patterns. Reducing ownership fragmentation of existing and proposed reserves on federal lands would improve management of lands for multiple uses and is well recognized (NRC 1993). Many mechanisms (e.g., conservation easements, land trusts, land trades, and dedications) have been attempted for a variety of purposes, sometimes to increase the representation of old-growth forest types that are underrepresented on the landscape. But land trades and similar mechanisms often are controversial, and many groups believe the government is not receiving equitable treatment in those transactions.

The major challenges for forest ecosystem management in the Pacific Northwest lie less in resolving the problems of the past than in anticipating the changes of the future. Development of the Pacific Northwest during this century has relied heavily on the array of resources—timber, domestic water, salmon, and recreation—provided by federal forests. Natural resources will play a prominent role in the future as well—the 1994 population is expected to have doubled by 2030.

3
OLD-GROWTH FORESTS

INTRODUCTION

Issues surrounding old-growth forests are at the very center of discussions about forest management in the Pacific Northwest. Terms such as "overmature," "late successional," "ancient forest," "forest primeval," as well as "old growth," often are used interchangeably, adding to the confusion of these discussions. Furthermore, public perceptions of what an old-growth forest is might not jibe with quantitative definitions of forest scientists and managers (Ribe 1989).

This chapter describes the attributes that characterize old-growth forests. Variations among forest types are described with respect to those attributes, as well as variations in the age at which forests acquire particular characteristics after disturbance. The characteristics impart several unique ecological features of old-growth forests, such as complexity and high biodiversity, low susceptibility to disturbance, and mesic microclimate.

WHAT IS VEGETATIVE SUCCESSION?

Succession implies structural and compositional change in the species that dominate a plant community (and frequently an animal community as well). Fire, wind, disease, and other disturbance processes of varying intensity and frequency select for adaptions in a landscape's biota, making some species more resistant to and other species more resilient to effects of disturbance; some species have evolved to become dependent on disturbance. A landscape can be viewed as a collection of patches of varying size and undergoing changes influenced by disturbance, as well as by the patches that surround it, and at any time, different parts of a forested landscape will be at different successional stages. A Pacific Northwest forest could comprise an evenly aged stand of red alder that eventually is replaced by Douglas-fir, which might then be replaced by hemlock or cedar.

WHAT IS AN OLD-GROWTH FOREST?

Late-successional, or late-seral, has commonly referred to forests in which shade-tolerant tree species, such as western hemlock and grand fir, begin to attain dominance (Spurr and Barnes 1973). FEMAT (1993) defined late-successional quite differently, as the period from first merchantibility to culmination of mean annual increment. As traditionally defined (e.g., by Spurr and Barnes), late-successional conditions in Pacific Northwest forests occurred rarely, only after many years in the old-growth condition and in the absence of significant disturbances that maintained dominance of less shade-tolerant species (most commonly Douglas-fir or ponderosa pine). Under FEMAT's definition, however, late-successional has nothing to do with dominance by shade-tolerant species, but rather is a stage of development of all forests that occurs well before, rather than in the later stages of, old-growth conditions. Unless noted otherwise, our use of late-successional will follow Spurr and Barnes (1973).

In the absence of fire or other disturbance, Douglas-fir in moderately moist Westside forests is gradually replaced over many centuries by shade-tolerant species, most commonly western hemlock. However, because of intermittent fires, shade-tolerant species rarely replace Douglas-fir altogether (Agee 1993). Similarly, frequent fires maintained dominance by ponderosa pine throughout most of the low- and mid-elevation forests in the interior of the Pacific Northwest; hence, forests dominated by shade-tolerant species were a minor component of the region. Late-successional forests have been more common at higher elevations in interior Oregon and Washington.

Old-growth forests are forests that have accumulated specific characteristics related to tree size, canopy structure, snags and woody debris, and plant associations. Ecological characteristics of old-growth forests emerge through the processes of succession. Certain features—presence of large, old trees, multilayered canopies, forest gaps, snags, woody debris, and a particular set of species that occur primarily in old-growth forests—do not appear simultaneously, nor at a fixed time in stand development. Old-growth forests support assemblages of plants and animals, environmental conditions, and ecological processes that are not found in younger forests (younger than 150-250 years) or in small patches of large, old trees. Specific attributes of old-growth forests develop through forest succession until the collective properties of an older forest are evident.

The U.S. Forest Service (USFS) Old-Growth Definition Task Group (1986) defined old-growth forests as the third of three basic stages in forest development. These forest stages are young, mature, and old; or, as sometimes distinguished by foresters, immature, mature, and overmature. In Douglas-fir forests of the Pacific Northwest, maturation typically occurs at 80 to 110 years. The mature forest represents a relatively stable stage with substantial continued growth and biomass accumulation, albeit at a slower rate than in the young forest. Transition from the mature to the old-growth stage is gradual. Douglas-fir stands do not begin to show the characteristics usually associated with old-growth until they are 175 to 200 years old.

> *Old-growth forests are forests that have accumulated specific characteristics related to tree size, canopy structure, snags and woody debris, and plant associations.*

Ecological characteristics of old-growth forests vary from one forest type to another (Tables 3-1 and 3-2) (Franklin and Spies, 1991a; Spies and Franklin, 1991), and therefore, no single definition of old growth is appropriate. Increasingly, definitions rely on indexes of successional development based on multiple forest characteristics (e.g., Spies 1991). USFS interim definitions for all types involve specific values or states for five criteria—number of large, old trees; variation in tree diameter; degree of tree decadence; amount of large, dead wood; and characteristics of the canopy architecture (USFS 1993b; Williams 1992).

- *Number of large, old trees.* The minimum density of large, old trees necessary for a stand to qualify as old growth varies from 16 to 50 per

TABLE 3-1. Biomass and Production by Ecosystem Components for Eastside and Westside Forests

Species	Location[1]	Age (years)	Above-ground biomass (Mg/ha)	Below-ground biomass (Mg/ha)	Detritus wood (Mg/ha)	Forest floor mass (Mg/ha)	Soil organic matter (Mg/ha)	Total live and dead biomass (Mg/ha)	Above-ground NPP[2] (Mg/ha/yr)	Total NPP (Mg/ha/yr)
WESTSIDE FORESTS										
Douglas fir	WA	36	173	33	6	16	112	340	-	-
	WA (SQ4)	22-73	132-304	-	-	-	-	-	5.1-9.3	-
	WA (SQ2)	40	453	-	-	-	-	-	13.7	-
	OR	50	301	67	12	7	88	475	-	-
	OR	70	464	-	-	29	251	-	-	-
	WA (SQ4)	40-70	170-256	30-44	-	27-28	50-70	277-398	9.1-8.9	12-12.5
	WA (SQ2)	40-70	306-379	56-67	-	23-23	70-120	455-589	11.1-15.8	13.7-18.8
	WA (SQ4)	150	763	141	-	30	50	984	13.5	17.8
	WA (SQ2)	150	812	154	-	20	110	1,096	9.6	11.5
	OR	90-110	661	-	-	-	-	-	12.7	10.9
	OR	150	865	-	-	-	-	-	10.5	-
	OR	450	718	153	215	51	113	1,250	8	10.9
	OR	26	193	38	90	22	770	1,113	32.2	34.9
Western hemlock/ sitka spruce	OR	121	916	187	21	34	776	1,934	22.8	25.4
Noble fir/ Douglas fir	OR	100-130	880	-	-	-	-	-	12.9	-
Mountain hemlock	WA	120-200	278	-	-	-	-	-	4.2	-

TABLE 3-1. (Continued)

Species	Location[1]	Age (years)	Above-ground biomass (Mg/ha)	Below-ground biomass (Mg/ha)	Detritus wood (Mg/ha)	Forest floor mass (Mg/ha)	Soil organic matter (Mg/ha)	Total live and dead biomass (Mg/ha)	Above-ground NPP[2] (Mg/ha/yr)	Total NPP (Mg/ha/yr)
Pacific silver fir	WA	130	453	-	-	-	-	-	4.9	-
	WA	180	446	138	240	150	273	1,247	2.3	14.5
	WA	300-350	542	-	-	-	-	-	-	-
EASTSIDE FORESTS										
Ponderosa pine	WA	120-200	136	-	-	-	-	-	-	-
Lodgepole pine/ western larch	WA	70	199	-	-	-	-	-	8.6	-
	WA	>180	339	-	-	-	-	-	-	-
Lodgepole pine	WA	65-70	-	-	-	-	-	-	2.6	-
Western white pine	IDAHO	100-250	265-330	-	-	-	-	-	4.7-10	-
	IDAHO	103	415-675	-	-	-	-	-	11.4-17.6	-
	IDAHO	103	488-794	-	-	-	-	-	13.1-20	-

[1] WA=Washington; OR=Oregon; SQ2=site quality 2; SQ4=site quality 4. Site quality is an index by which forests are grouped into classes based on past growth of trees on a site to predict future growth for a site. Site quality 2 is a high-productive site, while site quality 4 is a low-productive site.
[2] NPP=net primary production.
Source: Data from Vogt 1987, Gower et al. 1989, Vogt 1991, Vogt et al. 1996.

TABLE 3-2. Major Tree Species and Several of Their Dimensions in the Pacific Northwest

Species	Occurrence[1]	Fire tolerance[2]	Shade tolerance[3]	Modal (maximum age)	Number mycorrhizal species	Major abiotic and biotic constraints[4]
WESTSIDE FORESTS						
Abies amabilis (Pacific silver fir)	OR, WA	-	VT	350 (>500)	9	APH, T, W
Abies procera (Noble fir)	OR, WA	-	INT	400+ (>700)	1	NTM
Picea sitchensis (Sitka spruce)	CA, OR, WA	-	T	500 (>750)	21	NTM, W
Sequoia sempervirens (Redwood)	CA, OR	0 (young), + (mature)	T	1,000+ (>2,000)	?	M
Thuja plicata (Western redcedar)	CA, ID, OR, WA	-	T	200-500 (>1,200)	1 (vesicular-arbuscular)	APH, NTM
Tsuga heterophylla (Western hemlock)	CA, ID, OR, WA	-	VT	200+ (>500)	21	APH, BTR, DM, NTM, W
EASTSIDE AND WESTSIDE FORESTS						
Abies concolor (White fir)	CA, ID, OR, WA	- (young), 0 (mature)	VT	200 (300)	1	APH, DM, T, W
Abies grandis (Grand fir)	CA, ID, OR, WA	0	T	90-120 (250-280)	4	APH, BTR, M, T, W
Pinus contorta (Lodgepole pine)	ID, OR, WA	+	INT	200 (400)	25	BB, DM, M, W
Pinus monticola (Western white pine)	CA, ID, OR, WA	0	I	200 (500)	19	API, BB, DM, M, RR, T
Tsuga mertensiana (Mountain hemlock)	OR, WA	-	T	400+ (>800)	2	T, W

TABLE 3-2. (Continued)

Species	Occurrence[1]	Fire tolerance[2]	Shade tolerance[3]	Modal (maximum age)	Number mycorrhizal species	Major abiotic and biotic constraints[4]
Larix occidentalis (Western larch)	ID, OR, WA	0 (young), + (mature)	INT	300+ (>700)	19	APN, DM, IP, M, RR
Picea engelmanni (Engelmann spruce)	CA, ID, OR, WA	-	T-VT	300+ (>500)	6	M, RR, W
Pinus ponderosa (Ponderosa pine)	ID, OR, WA	- (young), + (mature)	INT	200+ (500)	19	API, DM, BB, M, T

[1]CA=California; ID=Idaho; OR=Oregon; WA=Washington.
[2]fire tolerance scale = + (high); - (detrimental); 0 (intermediate).
[3]shade tolerance scale = VT (very tolerant); T (tolerant); I (intermediate with seedlings more tolerant but mature trees less); INT (intolerant).
[4]major environmental constraint to growth = APH (highly sensitive to air pollution); API (intermediate sensitivity to air pollution); APN (not sensitive to air pollution); BB (bark beetles); BTR (butt and trunk rot); DM (dwarf mistletoe problems); IP (insect pest problems); M (moisture levels to low); NTM (no low temperature or moisture limits to growth); RR (root and butt rot); T (low temperatures, spring frosts); W (wind)
Source: Data from Trappe 1962; Hepting 1971; Franklin and Dyrness 1973; Franklin, 1979; Vogt et al. 1981; Molina and Trappe 1982; Parke et al. 1983.

ha, depending on forest type, with minimum sizes ranging from 52 cm diameter at breast height (dbh) on less-productive sites (generally in the interior) to 92 cm dbh on more-productive sites west of the Cascade crest. Old-growth forest might have 2 to 3 times that density of large, old trees. Minimum ages for these large dominant trees range from 150 years for the major interior forest types to 200 years for forests in western Oregon and Washington.

Old-growth Douglas-fir and ponderosa pine stands typically contain trees some 700 years old, and in some cases, more than 1,000 years. Old-growth Douglas-fir in western Oregon and Washington most commonly range from 350 to 700 years of age (Franklin et al. 1981). Under some circumstances, forests younger than 150-200 years produce trees that meet minimum size requirements, but those trees do not have characteristics of old-growth trees, such as thick, deeply incised bark and various manifestations of decadence as discussed below (Spies and Franklin 1991).

- *Variation in tree diameter.* Variation in tree diameters is greater in old-growth forests than in younger forests. For example, in the western hemlock zone of western Oregon and Washington, the standard deviation of tree diameters in stands 200 years or older is 2 to 3 times that of younger stands. In many forest types, that difference in diameter reflects the increasing abundance of shade-tolerant tree species in the understory as forests age. In other cases, such as ponderosa pine, it is due to small patches of young pine regenerating within the old-growth matrix.

- *Tree decadence.* In western Oregon and Washington, stands of Douglas-fir that are considered old growth have greater numbers of trees with broken tops, excavated bole cavities, root collar cavities, and bark resinosis than either young or mature stands (Spies and Franklin 1991). Those characteristics are typical of old trees throughout the region; old grand fir trees, for example, are commonly infected with a heart rot called Indian Paint fungus (*Echinodontium tinctorium*).

- *Presence of large dead wood.* Large, standing snags and fallen tree boles typify all types of old-growth forests. On average, 25-35% of the standing boles in old-growth Douglas-fir stands are snags comparable in size to living trees, with more than half of the snags larger than 50 cm diameter (Franklin and Spies 1991b). Large logs are also common on the forest floor (termed "down wood") in Douglas-fir old growth forests, as well as in high-elevation forest types. Down wood occurs in dry forest types, such as ponderosa pine, but less commonly than in more mesic forests. Snags and logs are important structural features of old-growth forests in providing wildlife habitat (Maser et al. 1979; Thomas et al. 1979; Harmon et al. 1986). In the Blue Mountains, for example, 39 bird and 23 mammal species use snags, and 179 vertebrate species make at least some use of down wood (Thomas et al. 1979; Maser et al. 1979).

Young, naturally established stands also can have large dead wood as legacies from the previous stand. Hence, presence of large, dead wood does not distinguish an old-growth forest as reliably among natural stands of different ages as does the number of large trees (Franklin and Spies 1991b; Spies and Franklin 1991).

- *Canopy architecture.* Large variation in tree diameters in old-growth forests is accompanied by a high degree of structural complexity in the forest canopy. Old-growth forests contain multiple tree-canopy layers (in addition to herb and shrub layers), a feature common to all types except ponderosa pine, lodgepole pine, and Eastside Douglas-fir. That layering reflects the growth of saplings (mostly shade-tolerant trees) into midcanopy strata as stands age. Old-growth forests in western Oregon and Washington also tend to have greater shrub and herb cover than younger stands (Spies and Franklin 1991). That is largely a function of overstory canopy density, inasmuch as open-grown younger stands generally have abundant shrub cover. Frequent ground fires retard understory growth in ponderosa pine and Eastside Douglas-fir forests.

A typical old-growth forest also has areas with little or no understory, resulting in a patchy spatial structure. Patchiness is a distinguishing characteristic of old-growth ponderosa pine, in which small islands (roughly 0.25-0.5 ha) of regenerating pine occur scattered through a matrix of older trees. Viewed from above, the pattern of canopy cover in old-growth stands is distinct from younger stands. Cohen et al. (1990) used low-altitude remote images from the central Oregon Cascades to calculate semivariance, a geostatistical technique that quantifies average patterns of variance. Different age classes of forests were easily distinguished visually using the red bands of the spectrum. Nel et al. (1994) assessed the value of remote imagery of canopy reflection to identify old growth in spruce-fir forests of the interior West. They found the same general patterns as Cohen et al. (1990) but concluded that although remote imagery was a useful guide to old-growth stands, it was not sufficient to identify old-growth stands with certainty.

Time Required for Old-Growth Development

Development of old-growth characteristics forests is progressive and varies among forest types. Some characteristics first appear about a century after a disturbance that destroys a forest stand (Table 3-3). Across the region, many old-growth characteristics develop during the second century of stand development, but forests do not typically display all of the properties described above before they are 200 years old.

Rates of succession differ from site to site depending on environmental conditions, nutrient and moisture availability, and residual forest components from the previous stand. To deal with the spatial and temporal variability in succession, Franklin and Spies (1991b) developed a continuously varying index of old growth based on the following five criteria for naturally established stands in the Oregon coast range, Cascades, and the southern Washington Cascades:

- density of large trees (e.g., more than 80cm Dbh)
- density of shade-tolerant trees
- amount of crown decadence (e.g., broken tops and multiple tops)
- density of large snags
- log biomass (using 60 Mg per ha as the base value)

On the Westside, Douglas-fir stands younger than 200 years generally have a low old-growth index (Franklin and Spies 1991b). Forests in that region have been classified as old growth at ages ranging from 150-250 years (Franklin and Spies 1991b; Johnson et al. 1991; Bonnicksen 1993; FEMAT 1993; Bolsinger and Waddell 1993). There is, however, considerable change in old-growth features across that age span. In the Franklin and Spies (1991b) study region, the old-growth index averaged a little more than 2.0 for stands younger than 100 years (range, 0-6.0), 3.0 for stands between 100 and 300 years (range, 0.3-5.0), and 6.0 for stands older than 300 years (range, 1.0-10.0) (Spies and Franklin 1991).

Old-Growth Landscapes

Ideas about how forests developed following major natural disturbances

TABLE 3-3. Standard of Old-Growth Characteristics for Different Forest Types in Oregon and Washington

WESTSIDE FORESTS	Live Trees Main Canopy DBH*	TPA*	AGE*	Variation in Tree Diameter (Yes or No)	Tree Decadence TPA	Tree Canopy Layers Number	Dead Trees Standing DBH	TPA	Down Diameter	Pieces
Douglas-fir										
High Sites (1,2,3)	37" (94cm)	8 (20tph)	190	Yes	No	2	13" (33cm)	1 (3tph)	24" (61cm)	4 (10/ha)
Medium Sites (4)	34" (86cm)	9 (22tph)	205	Yes	Yes	2	15" (38cm)	1 (3tph)	24" (61cm)	4 (10/ha)
Low sites (5)	24" (61cm)	10 (25tph)	200	Yes	Yes	2	17" (43cm)	1 (3tph)	24" (61cm)	4 (10/ha)
Western Hemlock										
Site Class 1	42" (107cm)	8 (20tph)	200	Yes	Yes	2	20" (51cm)	4 (10tph)	12" (30cm)	69 (170/ha)
Site Class 2	35" (89cm)	8 (20tph)	200	Yes	Yes	2	20" (51cm)	4 (10tph)	12" (30cm)	37 (91/ha)
Site Class 3	31" (79cm)	8 (20tph)	200	Yes	Yes	2	20" (51cm)	4 (10tph)	10" (25cm)	15 (37/ha)
Site Class 4 & 5	21" (53cm)	8 (20tph)	200	Yes	Yes	2	20" (51cm)	4 (10tph)	8" (20cm)	29 (72/ha)

Site										
Pacific Silver Fir										
Sites 2 & 3	26" (66cm)	6 (15tph)	180	Yes	Yes	2	22" (56cm)	6 (15tph)	24" (61cm)	4 (10tph)
Site 4	25" (64cm)	7 (17tph)	200	Yes	Yes	2	22" (56cm)	4 (10tph)	24" (61cm)	4 (10tph)
Site 5	22" (56cm)	9 (22tph)	260	Yes	Yes	2	22" (56cm)	1 (2tph)	24" (61cm)	4 (10tph)
Site 6	22" (56cm)	1 (2tph)	360	Yes	Yes	2	22" (56cm)	12 (30tph)	24" (61cm)	4 (10tph)
Port Orford Cedar										
All Sites	32" (80cm)	16 (40tph)	240	Yes	No data	2	14" (36cm)	6 (15tph)	13" (33cm)	27 (68/ha)
Tanoak (Redwood)										
All Sites	32" (80cm)	8 (20tph)	240	Yes	No data	2	32" (80cm)	3 (7.5tph)	13" (33cm)	20 (50/ha)
EASTSIDE FORESTS										
White fir/Grand fir (Central Oregon)										
Low & Medium Sites	21" (53cm)	10 (25tph)	150	Yes	Yes	2	14" (36cm)	1 (2.5tph)	12" (30cm)	5 (13/ha)
High Sites	21" (53cm)	15 (38tph)	150	Yes	Yes	2	14" (36cm)	1 (2.5tph)	12" (30cm)	5 (13/ha)

TABLE 3-3. (Continued)

	Live Trees Main Canopy			Variation in Tree Diameter	Tree Decadence	Tree Canopy Layers	Dead Trees Standing		Dead Trees Down	
White fir/Grand fir (Blue Mountains)										
Low & Medium Sites	21" (53cm)	10 (25tph)	150	Yes	Yes	2	14" (36cm)	1 (2.5tph)	12" (30cm)	5 (13/ha)
High Sites	21" (53cm)	20 (50tph)	150	Yes	Yes	2	14" (36cm)	1 (2.5tph)	12" (30cm)	5 (13/ha)
Ponderosa Pine										
Low Sites	21" (53cm)	10 (25tph)	150	Yes	---	1	14" (36cm)	3 (7.5tph)	---	0
	31"¹ (79cm)	2 (5tph)	200	Yes	Yes	1	Same	Same	Same	Same
Medium & High Sites	21" (53cm)	13 (33tph)	200	Yes	---	1	14" (36cm)	3 (7.5tph)	---	0
	31"¹ (79cm)	3 (8tph)		Yes	Yes	1	Same	Same	Same	Same
Douglas-fir										
All Sites	21" (53cm)	8 (20tph)	150	Yes	2 (5tph)	1	12" (30cm)	1 (2.5tph)	12" (30cm)	2 (5tph)

Lodgepole Pine (Sierra)

All Sites	12" (30cm)	60 (150tph)	120	Yes	No data	1	12" (30cm)	5 (13tph)	---	0

*Required minimums
(Metric equivalents in parenthesis)
[1]Very late seral (decadent) conditions

in the Douglas-fir region are changing significantly as a result of new information developed from historic stand reconstructions. The traditional view for coniferous forests (excepting ponderosa pine) is based on experience during the 20th century with stand development after logging (Oliver 1981; Oliver and Larson 1996). In that model, all trees composing the forest originate from the same disturbance, hence the stand is relatively even-aged. Trees grow into competition with one another and stands enter a self-thinning or "stem exclusion" phase, followed some time later by an "understory reinitiation stage" in which shade tolerant trees and shrubs become established. Over time, stands eventually attain old-growth status.

A growing number of studies suggest the traditional model does not accurately describe development of old growth in the Douglas-fir region (Tappeiner et al. 1997). Historic reconstructions performed to date, covering a wide area of western Oregon including the Cascade, Siskiyou, and Coast Range mountains, have consistently found old-growth stands consisted of multiple age classes of overstory trees spanning many tens of decades rather than a single age class (Franklin and Hemstrom 1981; Tappeiner et al. 1997). In their study of 10 old-growth stands in the Oregon Coast Range, Tappeiner et al (1997) concluded the oldest trees had established at low densities, with additional trees filling in over many years. They also found considerable patchiness within stands, some plots having a narrow age range of overstory trees and others having a wide range.

The picture now emerging is different and more complex than the traditional view, with considerable spatial and temporal heterogeneity, much less self-thinning than occurs in the even-aged stands created by clearcutting, an abundance of biological legacies, and intimate mixtures of older and younger trees intermingled both as individuals and in discrete patches. If a significant cover of hardwoods existed along with the conifers, which is suggested by the relative low conifer stocking, susceptibility to crown fires would have been low except under extreme weather conditions (Perry 1988a), and conifer diseases, a problem in many plantations, would have spread less readily than in densely stocked pure conifer stands (Manion 1981; Simard and Vyse 1994). The resulting old-growth landscape would have been dynamic but, because of a structure that tended to damp the spread of catastrophic disturbances, also relatively stable (Perry 1995a). While older trees, with their unique structure and habitat values, are a vital component (Bull et al 1992; FEMAT 1993; Peck and McCune 1997), the structure and functioning of old-growth landscapes emerges not from age alone, but from variable patterns of disturbance and succession within a more or less stable landscape matrix of older trees. The scale over which these dynamics played out historically, hence the characteristic spatial patterns of age classes and other structural attributes, varied among regions (e.g., from mesic to dry forests; SNEP 1996a,b), within watersheds (e.g., from southerly to northerly aspects and with elevation; Cissel et al. 1998), and over time in any one area due to the influence of varying macroclimate (Fryer and Johnson 1988; Swetnam 1993). As we discuss later, this view of a dynamic yet stable old-growth landscape serves as a basis for some of the new approaches to forest management being applied in the region.

Managed Forests and Old-Growth Characteristics

Managed forests can be thinned to produce large trees and structural heterogeneity at a relatively early age, especially in areas of high site quality. For example, a 75-year-old stand on the Black Rock State Forest (Oregon coast range), thinned to 125 trees per ha 30 years ago, now has an average tree diameter approaching the minimum old-growth requirement. But the degree to which such forests can be managed to mimic old-growth habitat and processes is unclear. Many managed forests either have no large dead wood or have much less than occurs in natural stands, which significantly reduces their habitat value for many animal species.

Some important characteristics of old trees are not related strictly to size and may be difficult to attain in trees younger than 175-200 years. For example, pileated woodpeckers, keystone species that excavate cavities used by numerous other bird and mammal species, prefer to roost in large grand fir trees extensively decayed by Indian paint fungus (Bull et al. 1992). The fungus is believed to enter through broken tops, which usually occur in older trees, and takes at least 20 years after entry to create sufficient decay for woodpecker use. Similarly, marbled murrelets prefer nest trees "in declining condition and (with) multiple defects (including) mistletoe blooms, unusual limb deformations, decadence, and tree damage" (Hamer and Nelson 1995). Alectroid lichens and cyanolichens — critical links in the food chain and, in the case of the latter, the nitrogen cycle — are much more diverse and abundant in old-growth than in young and mature stands, a phenomenon related in part at least to limited dispersal (Peck and McCune 1997), hence a matter of time. However, the time required for populations to recover may be shortened by the presence of hotspots of lichen diversity within young stands, especially old, remnant trees and hardwoods associated with gaps in the conifer canopy (Neitlich and McCune 1997).

Producing managed stands with large, old living and dead trees, dedicated to stay on site, is a primary objective of the relatively new silvicultural approach termed "variable retention" (Franklin et al. 1997) Such techniques attempt to mimic natural disturbances by preserving legacies such as decaying logs and accumulations of soil organic matter (Franklin 1993a). In time, practices that mimic natural disturbance patterns in managed stands might produce managed forests with at least some old-growth characteristics. However, the degree to which that is true can be determined only by extensive testing.

Stand Size

Edge effects (such as altered light conditions, animal activities, or species composition) can extend more than 100 m into a stand (Chen et al. 1992), which means that a 3-ha stand might exhibit edge effects throughout. The ecological relevance of edge effects depends on numerous factors. Small fragments are normally more vulnerable to drying (and hence to fire and wind damage) than large stands, but an old-growth fragment surrounded by mature forest presents a different situation than one surrounded by clearcuts. A fragment might be too small to provide habitat for some species, but it might be large enough for others. Furthermore, finding individuals of a given species in a fragment does not necessarily indicate the quality of the habitat. For example, marbled murrelets (*Brachyramphus marmoratus*) will nest in very small old-growth patches, but some evidence suggests that they are vulnerable to predators in such restricted habitats (Ralph et al. 1995).

The minimum size necessary for a stand to qualify as old growth has been subject to much debate. The Old-Growth Definition Task Group (1986) originally suggested 32 ha as a minimum, reasoning that, "patches of smaller size were not ... viable old-growth units because of their dominance by edge effects (penetration of external environmental influences) and vulnerability to major disturbances such as windthrow." However, that requirement was dropped because "minimum acreages for old-growth depended on management objectives and the nature of surrounding areas" (Old-Growth Definition Task Group 1986).

Fragmentation of intact old-growth stands increases vulnerability to disturbance and diminishes habitat values for at least some species; however, even very small old-growth fragments may retain some important biological values. The vulnerability and biological values of small patches depend heavily on the character of the landscape in which they are imbedded. Thus, no absolute statement regarding minimum size of old-growth stands can be made.

BIOLOGICAL FUNCTIONS OF
OLD-GROWTH AND LATE-SUCCESSIONAL FORESTS

The biological functioning of old-growth and late-successional forests is an important consideration in management of the Pacific Northwest terrestrial and aquatic ecosystems. To provide some understanding of the role of old-growth forests in the region, the discussion below centers on significant biological features and how biotic diversity of old-growth forests compares with that of younger forests.

Old-growth forests are biotically more complex than forests in earlier successional stages, and many species depend on the unique environmental features of old-growth forests for survival. Compared with younger forests, old-growth forests have a greater diversity of ecosystem components and specialized organisms and produce more food for some animal species (e.g., red-backed voles and spotted owls). They have a higher total amount of live and dead biomass (mainly due to the increased presence of coarse wood) and a higher amount of woody debris in streams and terrestrial areas. Old-growth forests are also less susceptible to large-scale disturbances and pest outbreaks, and they have a lower incidence of root-rot problems. Old-growth forests have unique microclimates and might have an effect on regional climate as well.

Species Diversity

The Scientific Analysis Team (SAT 1993) listed 80 terrestrial vertebrate species (16 amphibians, 38 birds, and 26 mammals) that were "closely associated" with old-growth stands in the range of the northern spotted owl, along with 99 invertebrate species and 111 vascular plant species. In his analysis of forests in western Oregon and Washington, Harris (1984) estimated that 118 terrestrial vertebrate species used old growth as primary habitat, 135 species used young forests (before canopy closure), and 90 species used mature forests. He also noted that "all of the species that meet their primary habitat requirements in...early stage forests find abundant habitat throughout the western Cascades and are generally common. Forty of the species finding primary habitat...in old-growth or mature forest cannot meet their habitat requirements outside this forest type."

Thomas et al. (1979) found somewhat different patterns in the Blue Mountains of eastern Oregon and Washington. In all forest types in the Blue Mountains, more terrestrial vertebrate species reproduced in mature and old-growth stands than in early-successional stands, but little difference was seen between old-growth and mature stands. The Blue Mountains and the Cascades have significantly different environments and hence, different patterns of forest development.

Arthropods and other small or inconspicuous organisms account for the bulk of diversity but have been largely overlooked until recently (Parsons et al. 1991), despite their importance in terms of numbers, species diversity, and functional importance. Lattin (1990) estimated that 8,000 arthropod species are found in the H.J. Andrews Experimental Forest in the central Cascades, compared with 143 vertebrate species and 460 vascular plant species. Many of those arthropods live either in canopies or soils, the least studied subsystems in forests. Schowalter (2000) compiled data from 9 forest, grassland, desert, and marsh ecosystems where extensive species inventories are available to show that arthropods commonly account for at least 70-90% of all species present.

The few studies that have been completed found striking differences between old-growth and younger forests in epiphyte, arthropod, and lichen communities. For example, one of the distinguishing characteristics of old-growth Douglas-fir forests is the abundance of epiphytic plants (mosses and lichens), especially the nitrogen-fixing lichen *Lobaria oregona*. *Lobaria* occurs in younger stands but not nearly as abundantly as in old-growth forests (Franklin et al. 1981).

Arthropod communities in old-growth canopies are significantly more diverse than in young plantations. Schowalter (1989, 1995) measured more than 70 species of arthropods associated with Douglas-fir and western hemlock foliage in old-growth forests in the central Cascades and only 15 species associated with Douglas-fir in 7- to 11-year-old stands. Diversity was 5-6 times greater in old-growth stands than in younger stands, with some of the most striking differences occurring in the diversity of predatory arthropods, such as spiders (Schowalter 1995). Moreover, the structure of arthropod communities differed significantly between young and old forests. In the former, the biomass of phytophages (largely aphids) was 800% greater than that of predators (e.g., ants, wasps, and spiders), while in the latter, the biomass of plant eaters (largely defoliators) was only 20% greater than that of predators. That pattern suggests more effective internal controls over plant eaters in old-growth stands than in younger stands.

The dominance of hardwoods and shrubs in young forests in part reflects the suppression of conifers by insects and pathogens at this stage. Conifers become re-established after populations of insects and pathogens have been reduced because they have so few conifer hosts (Goheen and Hansen 1993). The pines, western larch, western red cedar, Engelmann spruce, and western hemlock have higher tree mortality associated with root disease in early stages of stand development (younger than 30 years), but mortality decreases thereafter (Hagle and Goheen 1988).

Indigenous insects and diseases might have important roles in stand development by mitigating the effects of biomass accumulation and competitive stress through natural pruning, thinning, and cycling nutrients. Old-growth forests generally have higher populations of predatory insects than younger forests; those insects might help maintain populations of herbivorous insects at lower levels than in young stands (Schowalter 1989, 1995). Management practices that focus on short rotations or plantations of single species result in an overall loss of predators. Pests also are better able to find their hosts in such managed forests, and the systems become more susceptible to insect outbreaks (Hagle and Schmitz 1993, Schowalter 1995).

Logs and Woody Debris

Much of the influence of old-growth forests on environmental conditions is conferred by their large persistent structures. Whereas younger or smaller trees decompose relatively quickly (Harmon et al. 1986), large boles of old trees contain terpenoid and phenolic compounds in the heartwood that inhibit decay and provide structure and resources for soil and aquatic systems for centuries after tree death and fall (Harmon et al. 1986; Schowalter et al. 1992; Schowalter et al. 1998; Progar 2000).

Logs decomposing on slopes stabilize soils, retain moisture during dry periods (often more than 200% of wood dry weight, especially after a period of decay has increased porosity), and provide organic matter and nutrients that are tapped by mycorrhizae and roots penetrating wood from surrounding plants (Harmon et al. 1986; Schowalter et al.1992). Logs falling into streams create the pool-and-riffle structure that contributes to aquatic biodiversity. Logs are essential components of salmon habitat, slowing erosion from upslope and minimizing scouring of streambeds that degrade salmon habitat.

Although amounts of woody debris that can function in conserving mycorrhizal inoculi are highest in late-successional forests (Vogt et al. 1995), the actual diversity and biomass of mycorrhizal fungi may peak in the early stages of stand development. Only limited data exist on the pattern of mycorrhizal development with stand age (Vogt et al. 1992), but existing information suggests that the number of mycorrhizal species and the associated sporocarp biomass peak at canopy closure. Thus, total diversity of mycorrhizal fungi may be more closely tied to the net primary production of a forest than to structural components such as coarse wood. In contrast, the diversity of certain groups, such as the truffle formers, is higher in old-growth stands than in young stands (Amaranthus et al. 1994).

The number of sites for mycorrhizal fungal colonization of root tips is maximal at canopy closure (Vogt et al. 1983). The peaking of mycorrhizal fungi in early states of stand development might indicate a carbon cost to plants to maintain mycorrhizal associations (Vogt et al. 1991); hence, less carbon might be available to sustain all plant parts after that stage (Grier and Logan 1977). All Pacific Northwest forest tree species have a good complement of different mycorrhizal species that are capable of colonizing tree-root systems (e.g., almost 1,000 species of mycorrhizal associates have been reported for Douglas-fir (Perry et al. 1992)). Many of the trees share similar species of mycorrhizal associations.

Insects and pathogens are instrumental in directing succession through their selection of plant species. Although

Old-growth forests are more resistant to crown fires than are younger forests....

tree characteristics have been emphasized in most studies of succession, insects, pathogens, and other taxa influence seed production and dispersal, tree growth and survival, nutrient cycling, and soil-fertility patterns, and therefore affect the rate and direction of successional transformation. For example, several insects and root pathogens that kill Douglas-fir trees are instrumental in accelerating the transition to hemlock or cedar forest but also might provide litter accumulation sufficient to fuel a stand replacement fire (Goheen and Hansen 1993). Insects and root diseases responsible for stand replacement also can retard germination and growth of young conifers after disturbances.

Susceptibility to Disturbance

Old-growth forests are more resistant to crown fires than are younger forests, perhaps because of high humidity and litter moisture (Perry 1988a; Franklin et al. 1989; Chen et al. 1993). The hardwood and shrub species in the old-growth understory also appear to inhibit fire spread and to protect interspersed conifers. Young, evenly aged conifer forests are the most flammable and are particularly vulnerable to reburns (Agee 1993). Old-growth forests might be most vulnerable to fire where adjacent younger, drier, and more flammable forests provide the necessary heat and fuels to carry flames into the forest canopy. However, the heterogeneous structure of old-growth forests and the water-saturated logs provide barriers to fire spread, allow trees to survive, and provide open spaces for growth of understory (Perry 1988a; Palazzi et al. 1992).

AQUATIC ECOSYSTEMS

Streamside disturbance and flooding have important impacts on virtually all components of aquatic ecosystems (Reiter and Beschta 1995); however, no component has received more attention than salmon and trout. Seven species of salmon exist in the Pacific Ocean, and five occur on the North American continent: chinook (*Oncorhynchus tschawytscha*), coho (*O. kisutch*), chum (*O. keta*), pink (*O. gorbuscha*), and sockeye salmon (*O. nerka*). In addition, there are anadromous trout, steelhead or rainbow trout (*O. mykiss*), coastal cutthroat trout (*O. clarki*), and Dolly Varden trout (*Salvelinus malma*). Those salmonid species do not obligately require old-growth forests for survival, but they did evolve across a geographic range that closely overlaps that of the Northwestern coniferous forests.

Anadromous salmon and trout spawn in freshwater streams. Fry and juveniles rear in streams and rivers (sockeye in lakes), migrate to the ocean, spend varying amounts of time (depending on species and stocks), return as adults to their natal streams, spawn, and die. All Pacific salmon die after spawning, but a small fraction of anadromous trout are capable of repeated ocean migration and spawning. All eight species of the anadromous salmon and trout spend a portion of their lives in freshwater habitats in forested areas of the Pacific Northwest. As a result, their survival and production are closely linked to the forest ecosystem and are influenced by changes caused by forest practices (NRC 1996).

Disturbance, which is a critical feature of streams and rivers, strongly influences the survival of salmon and trout. Floods are a natural and essential component of rivers. Streams are shaped by floods, and rivers are more productive after flooding. Although productivity might decline immediately after flooding, flooding in streams creates pools, cleans gravels, and delivers dissolved and particulate nutrients. Refuges, such as deep pools, boulders and logs, off-channel habitats on floodplains, and stems and roots of streamside forests, are required for aquatic organisms to survive frequent disturbances. Over short- and long-term scales, old-growth forests along streams and floodplains create sizes and amounts of woody debris that cannot be provided by younger forests. Floodplain habitats, large woody debris, and pool habitats have declined substantially in recent years, and conversion of old-growth forests to younger stands is one of the causes of habitat losses related to the decline of Pacific salmon (NRC 1996). Many processes mediated by old-growth trees can be provided in riparian reserves or stream-management zones, but riparian management must be integrated with watershed conditions and land-use practices.

EXTENT AND STATUS OF OLD-GROWTH FORESTS

Differences in estimates of the past and present extent of old-growth forest in the Pacific Northwest reflect differences in the definitions used in each study: the time frame (e.g., presettlement, early settlement, pre-World War II, or current), the geographic area (e.g., states, Eastside forests, or Westside forests), land-use types (e.g., forest lands only or combinations of forest, agricultural, residential, and urban lands), and type of ownership (e.g., public, private, or national park). Eight studies have attempted to evaluate the current extent of old-growth forest in the Pacific Northwest, and eight studies have attempted to reconstruct the past distribution (Table 3-4). When allowances are made for the factors above, the study results are much the same.

Regional patterns of forest age classes and structure before logging resulted from the frequency and severity of natural disturbances, primarily fire, and to a lesser degree, wind, insects, and pathogens. The natural fire regime is closely linked to climate and, as a result, historic patterns of forest succession have varied within the region.

Estimates of the extent of old-growth before logging in the Douglas-fir

TABLE 3-4. Comparison of Studies of the Historical Extent of Old-Growth Forests in the Pacific Northwest

Study	Timeframe	Region	Landbase	Extent
Bonnicksen 1993	Pre-1850	W OR, W WA	Public and private forest land	42%-77%
Booth 1991	Pre-1850	W OR, W WA	Public and private forest land	62%
Teensma et al. 1991	1890	Coastal OR	Public and private forest land	46%
Morrison 1991	Pre-1850	W OR, W WA	Public and private forest land	66%
Morrison and Swanson 1990	1200-1990	Central Cascades, OR	Public forest land only	25%-49%
Andrews and Cowlin 1940	Mid-1930s	W OR, W WA, NW CA	Public and private forest land	44%
Cowlin et al. 1942	1936	E OR, E WA (excludes 2 counties)	Public and private forest land	73%
Lehmkuhl et al. 1994	1932-1959	E OR, E WA	Public forest only (6 national forests)	Mean-8.8%
Bolsinger and Berger 1975	mid-1950	E OR	Public land only (Ochoco National Forest)	44%
Hemstrom and Franklin 1982	1980 (represents historical)	Mt. Rainer National Park	Public forest land only	65%

region west of the Cascades crest, inferred largely from fire histories, range from 5.6 million ha to nearly 8 million ha, or 54-70% of the commercial forest area (Andrews and Cowlin 1940; Franklin and Spies 1984; Norse 1990; Booth 1994). The first inventory of Pacific Northwest forests on the west side of the Cascades, conducted in the mid-1930s after a period of heavy logging in the lowlands and severe logging-related fires, recorded old-growth forests covering 4 million ha, or 44% of the commercial forest (Andrews and Cowlin 1940). In southwestern Oregon, where little logging had occurred by the 1930s, old-growth trees of various species accounted for 73% of the commercial forest (Cowlin et al. 1942). In the Oregon Cascades and eastern coast range, only 8% of forests had been cut by the mid-1930s, and 57% remained in old-growth stands. In western Washington, 18% of the commercial forest had been logged by the 1930s, leaving 45% in old-growth stands.

Most reconstructions of presettlement conditions estimate that old-growth forests covered 54-70% of the forest area in western Washington and Oregon. Timber harvest and development have reduced this to 13-18% (Table 3-5).

Using a landscape simulation model driven by climate change and its coupling to fire, Wimberly et al. (2000) estimated that old-growth forest coverage in the Oregon Coast Range varied from 25-75% during the past 3000 years. The earliest quantitative estimate for the Oregon coast range was that old-growth trees covered 33% of the commercial forest in the mid-1930s (Andrews and Cowlin 1940). By that time, a significant amount had been logged, and large areas of old-growth trees had been destroyed by rampant wildfires. If logged areas and the Tillamook burn are added to the 1930s inventory, estimates of old-growth stands in presettlement coast range forests rises to 47%, which is consistent with the Teensma et al. (1991) estimate of 46% in 1890. Slightly more than one-third of coast range forests were between 90 and 200 years old in the 1930s survey. Because of logging-related fires, estimates of fire patterns based on wildfires during the past century are likely to overestimate presettlement fire frequency and, thus, underestimate the original extent of old-growth forests.

At the time of the 1930s survey, an additional 2.5 million ha in western Oregon and Washington were 90- to 160-year-old stands that had been established by fire. About 50% of those stands had trees averaging 50 cm dbh or more and were beginning to take on old-growth characteristics. Thus, by the 1930s, despite extensive logging, 68% of commercial forest land in western Oregon and Washington remained in what FEMAT (1993) classified as "late-successional/old-growth" (stands 80 years of age or older). Based on that figure, when Euro-Americans

TABLE 3-5. Comparison of Studies of the Existing Extent of Old-Growth Forests in the Pacific Northwest

Study	Timeframe	Region	Landbase	Extent
Morrison 1991	Mid 1980s	W OR, W WA	Public and private forest land	13%
Lehmkuhl et al. 1994	1985-1990	E OR, E WA	Public forest only	Mean-4.5%
FEMAT 1993	1990	W OR, W WA, NW CA	Public forest only	<20%
Hann et al. 1994	1992	W MT	Public forest only (Beaverhead N.F.)	3%-21%
Bolsinger and Waddell 1993	1992	W OR, W WA, NW CA	Public and private forest lands	18%

arrived in the area in the 1800s, as much as 80% of the forests in western Oregon and Washington were older than 80 years and about two-thirds were older than 200 years.

Forests of interior Oregon and Washington were also dominated by old-growth stands. The first comprehensive survey of forest resources in eastern Oregon and Washington (excluding northeastern Washington) was completed in 1936, at which time "the area of commercial forest land [was] characterized by a high proportion of old-growth" (Cowlin et al. 1942). The 1936 survey found that old-growth forests of all types made up 89% of the sawlog-sized stands and 73% of all commercial forest in eastern Oregon and Washington. Nearly two-thirds of Eastside forest lands covered by the 1936 survey were dominated by ponderosa pine, which, even after a period of heavy cutting that began in the early 1920s, was still mostly old-growth. If adjustments are made for logging before 1936, the original low-and midelevation ponderosa pine forests were nearly 90% old growth. Those stands typically contained trees up to 60-70 in dbh with most of the stand volume concentrated in trees 20-44 in dbh (Cowlin et al. 1942).

Old-Growth Forests

In the late 1800s on what is now the Ochoco National Forest, surveyors recorded ponderosa pine at 93% of the section corners in all but the wettest forest types (north slopes above 5,000 feet in elevation).

Various tree species other than ponderosa pine dominated forests at higher elevations and on moist sites at midelevations in the interior. In the 1936 survey, those species were classified as either large or small rather than as old-growth or second-growth, which was the case with ponderosa pine and Douglas-fir. In 1936, 71-96% of species other than ponderosa pine were classified as large, which almost certainly would fit within current definitions of old-growth for these types. In 1936, 11% of Eastside forest lands were lodgepole pine, a pioneer tree species that colonizes sites after wildfire. Most of those stands were classified as medium-sized and probably were not old-growth.

No comprehensive early surveys exist for Idaho, western Montana, and extreme northeastern Washington. Forest types of central Idaho are quite similar to those in the Blue Mountains of Oregon and Washington, and historic patterns are unlikely to have differed substantially between the two areas. In contrast, forest types in northeastern Washington and northern Idaho are unlike other Eastside forests, having higher proportions of Douglas-fir and a representation of species such as western hemlock and western red cedar that are more typical of Westside forests.

Historically, northeastern Washington and northern Idaho had extensive stands of western white pine that were probably mostly old growth. These have either been logged or killed by white pine blister rust (an introduced pathogen). The moist western red cedar stands, common in northeastern Washington and northern Idaho, probably had a relatively large proportion of old growth, because fires were infrequent and of a low intensity that seldom killed large overstory trees.

The most extensive and recent analysis of the current extent of old-growth forest included all federal, state, and private forest lands in western Washington, western Oregon, and northern California (Bolsinger and Waddell 1993). Old-growth forests were estimated to occupy 18% of the existing forest lands in the area. That analysis was based on recent maps and interpretations by the array of institutions and ownerships that have been responding to the debate over old-growth stands during the past 2 decades.

Conditions of Eastside forests were reviewed by the Eastside Forests Scientific Panel (Henjum et al. 1994), which concluded that approximately 25% of the land on eight national forests in eastern Washington and Oregon was old-growth forest. When public and private lands are considered, that proportion decreases to less than 20%, which is consistent with the distribution of old growth estimated for Westside forests (Bolsinger and Waddell 1993).

The composition of Pacific Northwest forests has changed dramatically over the past 10,000 years since the last ice age. Thus, the distribution and amount of old-growth forests before settlement represents only one point along a continuum of natural forest change. In some prehistoric periods, old-growth coniferous forests made up less than the 60-70% of the forest that settlers encountered when they first came to the Northwest (Brubaker et al. 1992; Whitlock 1992). Regardless of the extent that old-growth forest might have increased and decreased naturally over thousands of years, the reduction of old-growth over the past century is a more abrupt change than the forests have undergone since the last ice age.

SUMMARY

Because the ecological characteristics of old-growth forests vary from one forest to another, no single definition of old growth is appropriate. However, as knowledge has progressed, various indexes of successional development have been developed to characterize forests. Old-growth forests are biotically complex, with some species depending on unique features of old growth to survive, and the biological functioning of old-growth and late-successional forests is important to management of terrestrial and aquatic ecosystems.

Fifty percent of Pacific Northwest land is forested. Depending on locality, late-successional and old-growth forests originally made up from 54-70% of the forests, but now they are only 10-18%. Harvest since 1850 has removed more than 80% of the late-successional and old-growth forests of the Pacific Northwest; nonetheless, more than 80% of the remaining old-growth forests occur in national forests.

4
THE STATUS AND FUNCTIONING OF PACIFIC NORTHWEST FORESTS

INTRODUCTION

In this chapter, the status, condition, and sustainability of the forested ecosystems and associated plant and animal species of the Pacific Northwest are reviewed and assessed. The effects of forest-use patterns and management practices on timber and nontimber species are evaluated, and the discussion includes an analysis of species that are threatened or endangered by forest cutting and habitat fragmentation. The chapter ends by considering the long-term effects of the loss of biological diversity on the stability and functioning of ecosystems in general. The findings presented in this chapter provide important bases for the assessment of forest management practices in Chapters 7 and 8.

FOREST CONDITION

General Criteria of Condition

A system that is in good condition is one that retains its basic structures and processes (Rapport 1989). The condition of an ecosystem represents more than the absence of disease. It is also the ability to resist or recover quickly from environmental stressors. Because ecosystems seldom have clear boundaries, ecological condition spans spatial scales. The structure of landscapes, for example, shapes processes (e.g., hydrology and propagation of disturbances) that influence the integrity of stands and streams. The integrity of streams depends additionally on the integrity of riparian forests and of upslope forests that control sediment inputs to the streams. Individual species modify processes in many ways that influence ecological health. One example is the regulatory role played by birds and predatory insects in consuming tree-feeding insects (e.g., Torgersen et al. 1990; Marquis and Whelan 1994). The natural enemies of insect pests require habitats such as large dead wood.

The traditional approach to assessing forest condition on the basis of appearance (an inventory) may not detect early changes in condition, an issue that was addressed well by the National Research Council in 1994 (NRC 1994). The committee recommended that assessment of ecosystem condition also should consider 1) the stability of soils and watersheds, 2) the integrity of nutrient cycles and energy flows, and 3) the functioning of ecological processes that facilitate recovery from damage. Those same factors underlie the health of forests. The authoring committee concluded that assessing rangeland condition by comparing the abundance and kinds of plants growing in an area with a benchmark plant community (a list of plants expected on rangeland in excellent condition) was inadequate because it did not ensure protection of processes critical to ecosystem sustainability.

Not all ecosystem types are represented extensively on public lands in the Pacific Northwest–examples include lowland floodplain forests, oak woodlands, and coastal tidal marshes. And checkerboard ownerships of public and private lands hinder effective assessment and management of forest ecosystem patterns and processes.

Measurement of key ecological and ecosystem processes is more able than an inventory to provide sensitive indicators of forest condition. Such processes include 1) rates of nutrient (especially nitrogen) capture (from soil or atmosphere) and fixation in biotic tissue; 2) rates of water flux needed to maintain cellular function and evapotranspiration rates (to prevent cellular cavitation and wilting); 3) rates of nutrient cycling and biotic processes that ensure adequate supplies of critical nutrients (especially nitrogen) or maintain balance among critical nutrients (such as C:N:P:K:Ca[1] ratios) and minimize nutrient leaching from the system; and 4) rate of development of soil and canopy characteristics that maintain favorable temperature and humidity, atmospheric intercep-

---

[1]Carbon:nitrogen:phosphorus:potassium:calcium.

tion, and control of water flow and erosion. Clearly, it is not practical to measure these in all or even a large number of stands. However, such measurements from a representative subset of stands could provide critical insight into the long-term trends in forest condition that result from alternative management practices. Maintenance of species assemblages and ecosystem processes both are important in measuring forest condition.

Ecosystem health problems in Pacific Northwest forests can be grouped into three general, interrelated categories: 1) increased vulnerability to insects, pathogens, fire, and drought; 2) extreme fragmentation or loss of habitat and of its biological diversity; and 3) soil degradation. Old-growth forests can be lost to fire, drought, or insects just as they are lost to chainsaws. Disturbance regimes that are too frequent and too severe degrade soils and increase sediment export to streams. Degraded soils grow new forests slowly and in some cases, not at all (Perry et al. 1989a).

The Role of Biological Diversity

Efforts to conserve biological diversity are based on the assumption that biodiversity has value. That value has been documented in numerous publications, including books by Ehrlich and Ehrlich (1981), Wilson (1988, 1992), and NRC (1999b). Some of the major points raised in those treatments are reviewed below, as are more recent findings, especially those that relate the functioning of ecosystems to their biodiversity (Johnson et al. 1996).

The number and genetic variability of plant, animal, and microbial species that live in a given location is called its biodiversity. Biodiversity also encompasses biologically mediated processes in a habitat. Forestry and other land-use practices are influencing the biodiversity of the entire Pacific Northwest region. However, biodiversity is also important on small scales. Practices that reduce the biodiversity of a 1,000-acre forest stand, for instance, can greatly affect its functioning even if those practices do not threaten any species with extinction. The loss of local biodiversity — i.e., the loss of species from a given habitat — can be of great ecological importance. Factors that lead to the loss of local biodiversity include the conversion of naturally regenerating old-growth forests into rapid-rotation forests dominated by a single tree species; the fragmentation of habitats via road building, agriculture, and clearcutting; changes in fire regimes; fertilization; and application of pesticides.

The loss of local biodiversity is a concern because accumulating evidence indicates that viable populations of indigenous species are important to the rates, seasonality, and direction of processes contributing to overall ecosystem functioning(Temple 1977; Bormann and Likens 1979; Franklin et al. 1989; Schowalter and Filip 1993). A few experiments have addressed the functional importance of particular species or species assemblages. Tilman and Downing (1994) manipulated plant diversity in a grassland ecosystem via nitrogen addition and found that primary productivity during a drought was significantly related to plant diversity. Productivity in the lowest diversity plots dropped to about one-tenth of what it had been before the drought, but in the most diverse plots it only dropped to one-half of its predrought level. The productivity of the more diverse plots was stabilized by drought-tolerant species compensating via increased growth for reduced productivity of drought-intolerant species. Because high diversity stands are more likely to have disturbance-resistant species in them, on average they should be more stable, i.e., more resistant, to disturbance. Schowalter and Turchin (1993) measured the growth of experimentally introduced southern pine beetle (*Dendroctonus frontalis*) populations in pine/hardwood forests in which tree diversity had been manipulated by reducing densities of pines or hardwoods. Only dense, single-species plantations of pines were conducive to population growth of this pine-killing insect; the presence of nonhost hardwoods or shrubby vegetation apparently interfered with discovery of interspersed or scattered hosts. In both of the studies above, primary productivity was more stable when plant diversity was higher. These results are supported by much nonexperimental research on factors contributing to pest outbreaks (Kareiva 1983; Schowalter et al. 1986; Hunter and Aarssen 1988).

Viable populations of indigenous species might be critical for maintenance of ecological processes in Pacific Northwest forests, but few studies have addressed the contributions of particular species or species assemblages to processes.

Many examples of the importance of biodiversity in ecosystem functioning are obvious. A critical role of organisms is in decomposition—the breakdown of organic structures into their physical elements, including energy. There is increasing evidence that ecosystems with higher levels of organismal diversity are better at carrying out productivity (e.g., Risch 1980; Courtney 1985; Ewel 1986; Ewel et al. 1991; Frank and McNaughton 1991; Naeem et al. 1994; Tilman et al. 1996) and nutrient retention (Tilman et al. 1996). Organisms create structures and communities that interact with and alter the physical world—such as forests and coral reefs—and that provide habitat for other organisms that influence additional processes. Organisms and biological structures have important influences on the hydrologic cycle (e.g., through condensation, interception, and evapotranspiration) and on geomorphic processes, such as erosion.

That biodiversity of ecosystems is linked to their functioning was first proposed by Elton (1958), further developed by Odum (1969), and then called into question by May (1973), Goodman (1975), and others. However, May and Goodman were addressing a different aspect of stability than Elton or Odum. Elton and Odum were referring to the stability of an entire ecosystem, whereas May and Goodman were referring to the stability of a single species within an ecosystem. May, for instance, demonstrated theoretically that the population size of a species is expected to be more stable (i.e., return to equilibrium more rapidly after a perturbation) if the species lives by itself than if it competes with many other species. His result is still considered robust. However, May did not explore the effect of a perturbation on the stability of total community biomass, ecosystem primary productivity, soil nutrient conservation, or other such ecosystem characteristics. Those characteristics, however, were the attributes of greater interest to Elton and Odum. A recent field study (Tilman 1996) has supported both Elton and May.

Long-term acclimation of ecosystems to changes in climate and other environmental variables is primarily dependent upon available biodiversity. Obviously, greater numbers of species and greater genetic variability among species provides for a larger number of biological building blocks for ecosystem adjustment and acclimation. Given ever-changing environments, the capacity to acclimate is central to the long-term sustainability of ecosystem processes. Such changes are obvious in the shifts of species' abundances documented in 1,000- to 10,000-year records obtained by studying pollen profiles in lake sediments.

Relatively unimportant species restricted to particular microsites during one climatic regime often become important and widespread as climate shifts (e.g., Anderson 1990) or as a disease pathogen invades a habitat (Davis et al. 1997). The reservoir of genetic diversity within individual species and populations is central to their ability to adapt to environmental change. In view of this, focus on genetically engineered genotypes of crop plants and forest trees has raised concern regarding the loss of genetic diversity that might be important to future conditions.

Ewel and co-workers (1991) experimentally established tropical successional sequences that differed in plant biodiversity. They found that more diverse communities were more nutrient conserving and more productive. Naeem et al. (1994) experimentally established British grassland communities that differed in their plant, herbivore, and decomposer diversity. They found that diversity led to significantly increased primary productivity as measured by the rate of photosynthesis. Diversity also significantly affected decomposition, nutrient retention, and vegetation structure. In reviewing these studies on the effects of biodiversity on ecosystem productivity and stability, Kareiva (1994) concluded that the loss of biodiversity leads to "less productive ecosystems, vulnerable to environmental perturbations, and plagued by declining soil fertility." This conclusion was supported by a field study in which the plant diversity of 147 prairie grassland plots was manipulated and found to affect directly total plant community productivity, nutrient use, and nutrient leaching loss (Tilman et al. 1996).

The redistribution of species across the globe is one of the most significant effects that humans have on ecosystems. The negative consequences of exotic species in natural and managed ecosystems demonstrates that the contribution of biological diversity to ecosystem functioning is not merely a function of the number of species present, but of their identities and evolutionary interrelations.

Resistance and Resilience

Biodiversity provides stability (resistance) and recovery (resilience) in the face of disturbances that disrupt important ecosystem processes. Resistance often results from complex linkages among organisms, such as food webs that provide alternate pathways for achieving particular flows of energy and nutrients. For example, the presence of numerous fungal species capable of forming mycorrhizae in a terrestrial ecosystem buffers the ecosystem against the loss of individual species and makes total loss of mycorrhizal function unlikely.

McNaughton (1977) explored theoretically the possible effects of biodiversity on ecosystem stability. He found that increased functional diversity within an ecosystem was expected to make an ecosystem more stable. Numerous studies have demonstrated that increased plant diversity helps stabilize primary production in response to climatic change. For instance, in 1977, a severe drought in eastern Europe caused greatly decreased plant growth and crop yields. Leps et al. (1982) compared the effects of the drought on the productivity of two Czechoslovakian fields, one of which had low plant-species richness and the other with high plant-species richness. The productivity of the species-poor field fell to 1/3 of its predrought level, whereas that of the species-rich field only fell to 2/3 of its predrought level. Although any comparison of two sites is open to alternative interpretations, many subsequent, better-replicated studies have found this same effect. For instance, Frank and McNaughton (1991) found that the most diverse Yellowstone grasslands were most stable in response to drought. Similarly, a long-term study of 207 plots in Minnesota grassland and savanna also included a period during which there was a major drought. Because the productivity and species composition of all plots was annually measured before, during, and after the 1987-1988 drought, it was possible to include extensive statistical controls for numerous potentially confounding variables when determining the effects of biodiversity on the drought resistance and resilience of these plots (Tilman and Downing 1994). Primary productivity in more diverse communities was most resistant to and recovered most rapidly from drought. Indeed, the least-diverse plots suffered a 4- to 8-fold greater loss of productivity than the most diverse plots, and recovered much more slowly from the drought than the most diverse plots (Figure 4-1). Diversity increased ecosystem resistance and resilience, because more diverse areas were more likely to contain some species that were drought resistant. Those species increased growth in response to the decreased abundances of their drought-sensitive competitors.

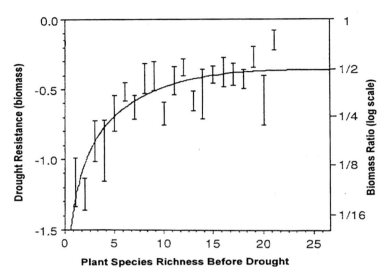

FIGURE 4-1. Effects of plant biodiversity on drought resistance of grasslands. Source: Adapted from Tilman and Downing 1994.

The significance of biodiversity in ecosystem resilience has been debated for many years, but evidence of its importance is now emerging from long-term research. Diversity-related resistance is particularly relevant to the management of agricultural and forest ecosystems to minimize the spread of species-specific pathogens and pest insects. And although single-species plantations result in high levels of production of specific products or resources, they also have a much higher risk of infestation than do more complex systems.

McNaughton (1985) presented data for the Serengeti savanna that demonstrated that areas with greater plant diversity were more resilient to natural grazing pressure, because ungulate grazers fed more selectively in areas with greater plant diversity. Plant species not consumed by grazers were able to increase rapidly in biomass once freed from competition with the species preferred by the grazers. Those compensatory increases tended to stabilize primary productivity in areas with greater plant biodiversity.

Landscape Change and Threats to Biodiversity

The long-term effects of ecosystem destruction and fragmentation are well known and well documented: habitat loss and fragmentation inevitably and unavoidably lead to species extinctions (e.g., MacArthur and Wilson 1967; Diamond 1972; Terborgh 1974; Ehrlich and Ehrlich 1981; Wilson 1988, 1992). Some rare species are extirpated when the only areas in which they live are destroyed. However, many other species are sufficiently harmed by habitat loss that their populations begin a slow decline that often ends in extinction (e.g., Diamond 1972; Terborgh 1974; Wilson 1992). Some species that survive the destruction of their prime habitat are left living in marginal habitats in which they cannot maintain viable populations. Such remnants are "sink" populations that are maintained only via continued immigration from neighboring viable habitats. The presence of transient species is thought to be the primary reason mainland sites of a given size contain more species than island habitats of the same size (MacArthur and Wilson 1967). Second habitat fragmentation isolates populations and decreases local population size. Small population size increases the chance of local extinction (May 1973), and isolation decreases the chance of recolonization by members of the same species. The long-term net effect is the eventual extinction of formerly abundant species (Tilman et al. 1994). Recent research postulates that the species most harmed by habitat fragmentation are the species that are best adapted (i.e., most specialized for the characteristic conditions) to the region and that they will undergo selective extinction in the remaining fragments of protected, undisturbed habitat (Tilman et al. 1994).

Island biogeographic theory (MacArthur and Wilson 1967; Simberloff 1984) predicts that the species richness that can be maintained within a particular region depends on the size of the region and on its degree of isolation from other regions. The close relationship between species richness and habitat area has been widely demonstrated (e.g., Tilman et al. 1994). Several factors contribute to species-area relationships. Habitat area alone affects species diversity because smaller areas support smaller populations that are more susceptible to extinction than larger populations in larger areas. Area is also a convenient surrogate variable for environmental characteristics correlated with it. For instance, as habitat size increases, the range of environmental conditions included within the habitat also might increase. In spatially homogeneous and heterogeneous regions, species richness increases with area, but the intercept and slope of that relationship depend on habitat heterogeneity (e.g., Simpson 1964; Pianka 1967; Greenstone 1984; Milne and Forman 1986). For instance, Simpson (1964) showed that mammal species per unit land area were greatest in areas of North America with the greatest topographic relief. The degree of isolation of an area, such as the distance of an island from the mainland or the fragmentation of forest patches by intervening agricultural activity, influences the rate at which species immigrate to that area. The dependence of species richness on area has been strongly supported by several studies from a wide range of ecosystems (e.g., Smith 1974; Power 1975; Rosenzweig 1975; Simberloff 1976; Molles 1978; Nilsson and Nilsson 1978; Rey 1981; Wells 1983; Brown and Gibson 1983; Malmquist 1985; Rydin and Borgegard 1988; Lomolino et al. 1989).

Over the past decade, the concept of metapopulation dynamics has been developed (see Hanski and Gilpin (1997)). Various authors have addressed the development of the metapopulation concept (Hanski and Simberloff 1997), effects of population fragmentation into more or less isolated demes on dispersal dynamics (Harrison and Taylor 1997), consequences for gene flow and genetic heterogeneity of the metapopulation (Hedrick and Gilpin 1997), capacity for recolonization following local extinction of isolated demes (Foley 1997; Thomas and Hanski 1997; Wiens 1997), ability of host-specialists to track their hosts in the shifting mosaic (Frank 1997), and application for conservation (Hanski and Simberloff 1997). As we would expect, populations maintain themselves as long as recolonization and population growth produce sufficient numbers of dispersing individuals to balance local extinction. Any population can be reduced or fragmented to the point where this balance no longer is maintained. Therefore, the most vulnerable taxa are, of course, those species that have relatively low reproductive rates and relatively low vagility, and thereby are less capable of maintaining gene flow and of recolonizing habitat islands in fragmented environments (Samways 1995). A number of studies from other regions have demonstrated reduced species diversity, influx of invasive species, and altered ecological function following fragmentation, e.g., Aizen and Feinsinger (1994), Bawa (1990), Klein (1989), Powell and Powell (1987), Punttila et al. (1994), Steffan-Dewenter and

Status and Functioning

Tscharntke (1997). Pollination and seed dispersal may be processes at particular risk, especially for increasingly isolated understory plants that depend on insect pollination and seed dispersal, as opposed to wind pollination, for successful recruitment (Aizen and Feinsinger 1994; Bawa 1990; Powell and Powell 1987; Steffan-Dewenter and Tscharntke 1994), a dependency shared by many understory plants in Pacific Northwest forests.

Other studies describe how habitat loss relates to the loss of biodiversity (Harris 1984; Simberloff 1984; Wilson 1988, 1992). The relationship between the amount of habitat exploited and the rate of extinction is not linear, nor are the extinctions resulting from habitat destruction usually instantaneous.

A recent experimental study of habitat fragmentation provides insights into some of its other effects (Kruess and Tscharntke 1994). Kruess and Tscharntke established local plant populations that differed in their degree of isolation and fragmentation. They observed that the plant-feeding insects that attack the plant species were equally good at colonizing plants, independent of the degree of plant population isolation. However, the predators and parasitoids of the insects were much less abundant in more isolated, fragmented populations. Plant-feeding insects in large, intact stands of this plant species were 2 to 5 times more likely to be attacked by predators and parasitoids than those in fragmented, isolated patches. Thus, habitat fragmentation, by preferentially harming predators and parasitoids, led to increased damage to plants from herbivores. Those results are consistent with Schowalter's (1995) results in Pacific Northwest forests.

Figure 4-2 illustrates the link between habitat loss and species extinctions. It is assumed that the relationship between the number of species (S) and area (A) can be approximated by the simple formula $S = cA^z$, where c is equivalent to the average number of species encountered in an area of unit size, and z is the fractional increase in species per fractional increase in area. The effect of habitat loss on regional species diversity is a sharply increasing function of the proportion of the habitat destroyed. The two curves shown in Figure 4-2—for z values of 0.1 and 0.35—span the known range of z values, and thus represent likely upper and lower bounds on the effects of a given degree of habitat loss on the eventual extinction of species. For instance, the destruction of 50% of a habitat should lead to the eventual loss of 7-22% of the existing species;

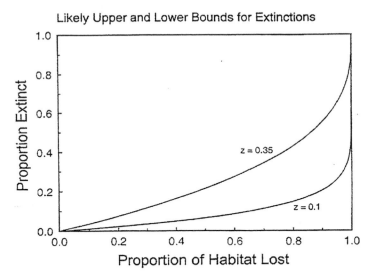

FIGURE 4-2. Effect of habitat loss on the extinction of species, derived from the species-area relationship (see Tilman et al. 1994).

destruction of 95% should lead to the extinction of 26-65% of the species. A small increase in the proportion of habitat destroyed would lead to many more extinctions in an already fragmented landscape than would a comparable increase in destruction of an intact landscape. For a virgin mainland habitat, z should be about 0.15, but should increase to as much as 0.35 as more of the habitat is destroyed and made more islandlike.

Much of the previous discussion is based on theory and data from ecosystems outside the Pacific Northwest. Based on these considerations and given that about 80-90% of the Pacific Northwest has already been subject to forest cutting, one might suggest that additional loss of old-growth forest to agriculture, sprawl or short-rotation silviculture might threaten an ever-larger number of species. The steeply rising curves in Figure 4-2 show that loss of an additional 1% of the original habitat when 90% has already been destroyed is expected to lead to the extinction of more than 4 times as many species as would a 1% loss in habitat when only 10% has been destroyed.

Certainly, no one would argue that habitat fragmentation or loss have

Status and Functioning

anything but negative effects on the viability of species populations. Furthermore, those effects are likely to increase disproportionately as total habitat area decreases and fragmentation increases. But several factors make it very difficult to make quantitative predictions of the extent of actual loss we might expect in the Pacific Northwest for some additional increment of old-growth forest loss (Rochelle et al. 1999). First, the analogy between a fragmented landscape and islands in an ocean is imperfect. The matrix of disturbed lands in the Pacific Northwest represents a complex collection of forested and nonforested lands that vary in their value as habitat and dispersal corridors. Second, relatively few species are entirely restricted to old-growth forests, making generalizations about the specific number of species jeopardized by an increment of old-growth loss impossible. Third, forests are constantly changing; cut-over forests do regrow and eventually reacquire old-growth features.

> *Pest outbreaks are not random events that threaten forest resources but rather, at least in part, are predictable responses to changes in forest condition; thus, they are valuable indicators of changes in forest condition.*

These caveats do not change the fact that continued cutting of old-growth forests and conversion of naturally regenerating forests to other conditions poses greater risks for the biodiversity of the Pacific Northwest now than it ever has. This is a matter of particular concern for those species that depend most on the habitat features (e.g., standing and downed woody debris) of old-growth forests.

Diseases and Pests

Pest outbreaks are not random events that threaten forest resources but rather, at least in part, are predictable responses to changes in forest condition; thus, they are valuable indicators of changes in forest condition. Outbreaks of insect and pathogen populations are affected by three general factors: 1) host condition; 2) host abundance and 3) host apparency, i.e., the ease with which an insect or pathogen can find a suitable tree species.

Host condition. Host condition indicates plant responses to environmental factors. Plants respond to stresses such as extreme temperatures, flooding, and moisture or nutrient limitation by adjusting their physiological processes to increase evapotranspiration or to conserve water and nutrients (Mattson 1980; Mattson and Haack 1987; Waring and Cobb 1992; Lorio 1993; Koricheva et al. 1998; Schowalter et al. 1999). Defensive chemicals, such as phenolics and terpenoids, which normally protect healthy plants from insects and pathogens, consume energy and nutrients that otherwise could be used for growth or to meet more immediate metabolic requirements associated with plant survival (Perry and Pitman 1983; Rhoades 1983; Strong et al. 1984; Becerra 1994; Harborne 1994). Stressed plants typically reduce production of such defensive chemicals (Rhoades 1983; Lorio 1993).

Plant-feeding insects and pathogens have a wide range of abilities to colonize host plants, depending on their tolerances or ability to detoxify plant defensive chemicals (Harborne 1994). Some insects and pathogens respond rapidly to small changes in concentrations of defensive chemicals in mildly stressed plants, whereas others can reach outbreak levels only if there has been a considerable reduction in plant defensive-chemical production (Lorio 1993). Some of the more virulent early colonists feeding on mildly stressed plants can exacerbate stress and predispose plants to other insects and pathogens (Schowalter 1985; Paine et al. 1993), leading to rapid decline in plant condition. For example, tree defoliation after infection by root diseases often weakens trees sufficiently to permit colonization by tree-killing bark insects (Goheen and Hansen 1993).

Host abundance. Host abundance strongly affects insect and pathogen populations. Insects and pathogens usually occur in scattered stressed or injured plants, where their effects are primarily to thin stands and accelerate successional transitions from shade-intolerant to shade-tolerant vegetation (Schowalter 1981). Insect and pathogen populations typically remain small overall because individuals dispersing in search of new hosts are vulnerable to a variety of mortality agents, including adverse weather conditions and predators. Most major outbreaks occur when insect and pathogen populations can grow on closely spaced and suitable hosts, a condition found most often in forest stands dominated by a single species, such as in forest plantations. Survival increases because more suitable hosts are available and easily reached (Risch 1980, 1981; Strong et al. 1984; Courtney 1985; Schowalter and Turchin 1993; Schowalter et al. 1986).

Host apparency. This term refers to the ease with which an insect or pathogen can find individuals of a suitable tree species. Insects adapted to feeding on particular plant species often use the defensive chemicals produced by their host plants and carried in the forest vapor stream (especially if plant injury increases chemical release to the airstream) as chemical beacons to locate host plants (Stanton 1983; Schowalter et al. 1986; Visser 1986). Chemolocation can be disrupted by nonhost plants that produce and emit other chemicals that do not attract or even repel insects or pathogens (Visser 1986). Thus, in vegetation containing a mixture of many plant species, host-finding by insects is difficult. In contrast, the strongly attractive vapor stream of single-species stands facilitate host-finding (Risch 1980, 1981; Stanton 1983; Strong et al. 1984; Courtney 1985; Schowalter 1986; Visser 1986).

As a result of host condition, abundance, and apparency operating interactively, single-species plantations are particularly vulnerable to pest outbreaks. Commercial tree species typically are rapidly growing species that easily become stressed by water or nutrient limitation (Lorio 1993). Stress, combined with abundance and apparency, greatly increase the likelihood of pest outbreaks in single-species plantations, especially in forests where these conditions persist for long periods (Kareiva 1983; Schowalter et al. 1986; Schowalter and Turchin 1993). The trend in silviculture toward planting genetically improved stocks might increase such risks by diminishing genetic variability in disease resistance.

Incidence of Pest and Disease Outbreaks

Susceptibility to pests and disease in the Pacific Northwest has increased primarily in Eastside forests. The change in condition has been created by altered tree species compositions since the late 1800's (Carlson and Lotan 1988; Hagle and Goheen 1988), aggravated by drought (Patterson 1992). Selective logging, especially removal of shade-intolerant species (e.g., western larch and ponderosa pine) has increased the abundance and distribution of drought-intolerant Douglas-fir and true firs, which have become susceptible targets of defoliators, bark beetles, and root pathogens (Hagle and Schmitz 1993).

Various agents in western forests—tree-feeding insects, pathogens, and fire—have positive and negative roles. For example, stem decay fungi are particularly important to cavity-nesting birds, because they soften heartwood and facilitate excavation of cavities in living trees (Bull and Partridge 1986). "Witches-brooms" formed by dwarf mistletoe (*Arceuthobium* spp.) provide shelter for owls and porcupines. Root rots create patchiness that diversifies plant species composition. At low levels, spruce budworm (*Choristoneura occidentalis*) kills weaker trees, thereby providing more resources for the more-tolerant survivors. Although some native pests occasionally kill trees over large areas, the structure of ecosystems and landscapes, along with the ecological relationships mediated by that structure, often dampens and absorbs the spread of disturbances, diseases, and pests (Perry et al. 1989b, 1994).

Some management practices for Westside forests were developed to minimize real or perceived pest problems. For instance, Port Orford cedar root rot (*Phytophthora lateralis*) has been limited by sanitation of logging equipment and forest closure. (The rot was introduced into high-elevation areas by vehicle movement from infected valley areas and subsequently transmitted by down-slope water flow.) Slash burning and other residue treatments have been prescribed to prevent spread of bark beetles and pathogenic fungi into crop trees. However, few of those insects or fungi were problems in old-growth forests, because species diversity and diversity of natural enemies limited pest abundance and spread (Schowalter 1989; Goheen and Hansen 1993; Schowalter 1995).

Lower tree species diversity in managed forests on the Eastside has increased the degree of disease development and severity on many sites and contributed to the higher mortality rates of trees than are evident on the Westside. For example, current annual mortality of Eastside Douglas-fir is estimated to be 4.1%, which is 4 times higher than its natural mortality rate; it is mainly due to fungal diseases (Hagle and Goheen 1988; Spies et al. 1990). Natural mortality is caused primarily by the thinning associated with stand development as trees increasingly compete for space and also to periodic beetle outbreaks (Rebertus et al. 1992). Hagle and Goheen (1988) suggested that Douglas-fir and the true firs are most susceptible to damage by root diseases in western Montana, northern Idaho, eastern Oregon, and eastern Washington and predicted chronic mortality of these species from root diseases would occur throughout the lives of the stands in which they are present.

Nutrient status of a site and disease also are causally associated. For example, *Armillaria* root rot problems are more severe in productive sites than in harsh, less-productive sites because of more intense logging on productive sites and the resultant increased true fir component (Hagle and Schmitz 1993).

Some insect outbreaks result from changing climatic conditions, such as droughts, and are especially severe on sites where dominance has shifted to tree species susceptible to insects and pests (Mattson and Haack 1987). For example, in west-central Idaho in the Cascade Ranger District, precipitation decreased 25% from 1985-1990. Mature Douglas-fir trees became stressed and were attacked by the Douglas-fir bark beetle (*Dendroctonus pseudotsugae*) (Patterson 1992). As a result, annual Douglas-fir beetle-related mortality increased from fewer than 1,000 trees killed to more than 15,000 in 1989 (Patterson 1992).

Two major factors underlie current disease and pest problems in forests east of the Cascades crest. First is the large-scale conversion of Eastside forests from a landscape dominated by a mosaic of old-growth pine to one dominated by younger forests with a high component of true firs and Douglas-fir. That has effectively converted the regional landscape from one that dampens and absorbs disturbances, such as fire and pest outbreaks, into one that magnifies them (Perry 1988b, 1994). This is a major reason for proposals to move landscapes of the Eastside back toward the historic dominance by large ponderosa pine (Agee 1993). The second factor is the introduction of foreign disturbances, i.e., stresses that have not been present historically, and against which the system has no effective defenses. Exotic insects and pathogens are obvious introductions, exemplified by the presence of nonindigenous white pine blister rust. And the potential for introducing and dispersing new exotic pest organisms could be exacerbated by U.S. importation of raw logs. Foreign stresses also include clearcutting and removing coarse woody debris (which reduce habitat for natural predators and parasites of pest insects and pathogens (Campbell and Torgersen 1982; Schowalter 1995), logging with heavy equipment (which compacts soils and diminishes ecosystem resilience), and disrupting the historic regime of frequent, gentle fires. Fire reduction and logging have been the major contributors to change in forest structure and landscape patterns (Agee 1993; Hagle and Schmitz 1993).

Considerable evidence from throughout the interior west indicates that spruce budworm outbreaks during this century have been more frequent, more widespread, of longer duration, and more lethal to trees than previously (Anderson et al. 1987; Swetnam and Lynch 1989; Wickman et al. 1992). Increasing populations of defoliating insects during the 20th century have resulted in large part from increases in true fir and Douglas-fir ingrowth in stands previously dominated by nonhost ponderosa pine and larch (Hagle and Schmitz 1993). Incidence of the root rot *Armillaria* might also have been increased by the spread of these tree species.

The basic ecological principle of host abundance discussed above is at work on the Eastside. True firs and Douglas-fir have become more abundant during the 20th century because early loggers took only ponderosa pine, leaving behind the less-valuable firs. Furthermore, the spread of firs was facilitated by fire suppression (Agee 1993). The resulting shift in tree species composition was dramatic. Eastside commercial forest land dominated by ponderosa pine has shrunk during the 20th century 75%-90%, depending on locale. Most of those forests are now dominated by Douglas-fir and true firs that once occurred primarily at the higher elevations.

Larvae of spruce budworm and tussock moth (*Orgyia pseudotsugata*), the principal outbreak defoliators in Eastside forests, are dispersed largely by wind currents. Larval survival, hence the rate at which an infestation spreads and the area it covers, depends (among other things) on the proportion of dispersing larvae that lands on a suitable host. That depends, in turn, on the relative proportion of host and nonhost tree species across the landscape (Schowalter et al. 1986; Perry 1988a; Schowalter and Turchin 1993). When Douglas-fir and true firs existed as forested islands on mountains surrounded by nonhost ponderosa pine at mid and low elevations, dispersal of defoliators was limited. The much greater expanse of firs today facilitates the spread of insects, a phenomenon that has been exacerbated by several additional factors. Very high stocking density in some stands (due to fire exclusion) stresses trees, reducing their ability to resist insect and pathogen attacks. Firs, less-drought tolerant than ponderosa pine, are particularly susceptible to drought stress in the lower elevation areas to which they have migrated since fire exclusion was adopted as a management strategy. Abnormally low precipitation during the 1980s probably contributed to the most recent spruce budworm infestation. Reduced habitat (especially coarse woody debris) for birds, ants, and other natural enemies might have also contributed to increased population growth and spread of defoliators (Torgersen et al. 1990).

In 1989, bark beetles and spruce budworm infestation was evident over 3.1 million acres in eastern Oregon. By 1991, more than 4.7 million acres was infested (Oregon Dept. of Forestry 1999). The area of active infestation has since declined, comprising 2 million acres in 1992, and measurements in spring of 1993 indicated that budworm populations were very low. Nonetheless, the potential for widespread infestations of budworm, tussock moth, and *Armillaria* will remain as long as current forest conditions persist.

Comparative studies would help to elucidate the numerous factors that relate to pathogen and insect outbreaks in different forests and to understand pathogen and insect prevalence.

STATUS OF OTHER PLANT SPECIES

In addition to their effects on timber species, land practices in the Pacific Northwest have influenced the stability and abundances of many other plant species of the region. The species that have been most immediately and greatly affected by past land use in the region are those species most closely associated with late successional and old-growth forests. Because most plant species abundances do not vary greatly among younger, mature, and old-growth forests in western Oregon and Washington (Spies 1991), few plant species have been directly endangered by the forest harvesting that has occurred. But that does not mean that their future is ensured. Indeed, habitat destruction and fragmentation might eventually threaten many common plant species of the region (Tilman et al. 1994). However, Pacific yew (*Taxus brevifolia*) and the lichen *Lobaria* spp., which have strong preferences for old-growth forests, are strongly affected by decreases in the spatial extent of old-growth forests (Spies 1991). In Idaho, Henderson et al. (1977) proposed 17 forest plants for federal listing under the Endangered Species Act

(ESA), including 11 for threatened status and 6 as endangered. In contrast, many more animal species have been adversely affected by land use and deforestation.

STATUS OF WILDLIFE

Habitat fragmentation, increased abundances of Douglas-fir and true fir, decreased fire frequency, the much greater abundances of early successional forests, soil erosion into streams, and construction of dams have affected many bird, mammal, fish, and amphibian species of the Pacific Northwest. Species that rely on old growth have frequently been greatly affected, and concern is growing for amphibians and reptiles, neotropical migrant passerine birds, endemic species, and other species that appear to be scarce but for which little is known. And although the ESA has focused attention on species such as the northern spotted owl and the grizzly bear (*Ursus arctos horribilis*), mandates predating the Multiple Use-Sustained Yield Act of 1960 pertain to national forests and address the need to maintain key ecological processes and biodiversity.

Paulson (1992) summarized distribution and status of the avifauna in the Pacific Northwest, including Idaho, western Montana, Alaska south of the tundra, and British Columbia. Of 380 species that occur regularly in the region, 327 use the area for breeding. An additional 110 species occur irregularly in the region and were not considered. The region has no endemic species, although seven species (trumpeter swan (*Cygnus buccinator*), Barrow's goldeneye (*Bucephala islandica*), red-breasted sapsucker (*Sphyrapicus rugar*), white-headed woodpecker (*Picoides albolarvatus*), chestnut-backed chickadee (*Parus rufescens*), varied thrush (*Ixoreus naevius*), and golden-crowned sparrow (*Zonotrichia atricapilla*)) are largely restricted to it. Vaux's swift (*Chaetura vauxi*), rufous hummingbird (*Selasphorus rufus*), Pacific-slope flycatcher (*Empidonax difficilis*), flammulated owl (*Otus flammeolus*), and Townsend's warbler (*Dendroica townsendi*) and hermit warbler (*Dendroica occidentalis*) breed primarily within the region and winter elsewhere. Breeding species richness (the number of species) is high in the Rocky Mountain portion of this region because of the variety of habitats and climates, with a peak in richness located in northwestern Montana (Cook 1969; Paulson 1992).

Species that have shown sharp declines in the Rocky Mountain region include the osprey (*Pandion haliaetus*), bald eagle (*Haliaeetus leucocephalus*), peregrine falcon (*Falco peregrinus*), willow flycatcher (*Empidonax traillii*), common nighthawk (*Chordeiles minor*), Lewis' woodpecker (*Melanerpes lewis*), olive-sided flycatcher (*Contopus cooperi*), western wood pewee (*Contopus sordidulus*), western bluebird (*Sialia mexicana*), mountain bluebird (*Sialia currucoides*), loggerhead shrike (*Lanius ludovicianus*), and red-eyed vireo (*Vireo olivaceus*) (Paulson 1992). Habitat loss on southern winter range and on breeding range, habitat fragmentation that predisposes interior forest species to increased brood parasitism from brown-headed cowbirds (*Molothrus ater*), and predation from other species are probable causes (Paulson 1992). Carter and Barker (1993) evaluated the status and reliability of data for these species from breeding bird surveys. Finch (1991) pointed out that status of neotropical migrants in western North America needs more evaluation.

For western Oregon (Meslow and Wight 1975) and the Rockies (Tobalske et al. 1991), logging changes species composition of forest passerines. Ground-nesting species and species that use openings and low-growing vegetation are found in recent clearcuts and early stages of regeneration. Foliage-foraging species and tree gleaners are less abundant in those areas. Conifer-tree-nesting species are least abundant, including golden-crowned kinglet (*Regulus satrapa*), Swainson's thrush (*Catharus ustulatus*), varied thrush, and Townsend's warbler.

McGarigal and McComb (1995) reported that 10 of 15 passerines were negatively affected by change in habitat area. Only 2 of the 15 species examined were affected by habitat configuration, and 4 species were positively associated with more fragmented habitats. Seven species were associated with landscapes containing more fragmented late seral forests than expected. These authors concluded that the relationship between late-seral forest area and bird abundance varied dramatically among species.

Raptors (hawks, owls, and eagles) pose special problems for habitat retention in managed coniferous forests in the region. McClelland et al. (1979) recommended that forest-management practices should minimize harassment to prevent nest sites and territories from being abandoned, prevent direct destruction of habitat, and provide selective cutting systems that remove only specific trees and ensure retention of snags and potential snags. Uncommon raptor species of Pacific Northwest coniferous forests include the flammulated owl and the boreal owl

(*Aegolius funereus*), which are associated with old-growth ponderosa pine forests and spruce-fir forests, respectively (Holt and Hillis 1987; Hayward et al. 1993). Goshawks (*Accipiter gentilis*) require protection of riparian habitat and protection of known active nest sites, which usually are found only in mature forest with canopy closure (Thomas et al. 1993).

Because much of the controversy has centered on spotted owl populations, much is known about that species. Noon and Biles (1990) examined population attributes of the species to see what factors are most responsible for population fluctuation. The spotted owl is a typical K-selected species (i.e., has a long generation time and small clutch sizes) with age at first reproduction being 2 years and with 2 eggs produced annually thereafter. Probabilities of survival to age 1 are low, but year-to-year probabilities become increasingly higher thereafter; maximum life is approximately 16 years. Noon and Biles (1990) demonstrated that spotted-owl populations are not sensitive to a wide range of changes in juvenile survival; the factor that most affects population dynamics is adult survival. Species that exhibit high adult survival and low juvenile survival are thought to evolve in habitat that favors adult longevity. Forests west of the Cascades crest and throughout much of eastern Oregon and Washington are generally more long lived and stable in composition than forests of the northern Rockies, which encourages adult owl survival. Relatively low fecundity precludes rapid repopulation after a decline, further suggesting adaptation to long-lived forests. Isolated habitats caused by logging, tree falls, or fire might not be recolonized rapidly, because survival of potential dispersers, which are typically young owls, is low.

Table 4-1 demonstrates that populations of spotted owls decreased from 1985-1993, and the rate of decrease accelerated in the 11 areas studied, consistent with earlier analyses of life-table data by Noon and Biles (1990), which showed a declining population level. The decrease was attributed (Lande 1988; Doak 1989; Thomas et al. 1990) to the loss of mature and old-growth forest habitat. Thomas et al. (1993) projected habitat declines at rates of 1.1–5.4% per year on intensive study areas.

Spotted-owl home ranges in Westside forests are typically 1,200-2,000 ha, of which 20-50% is old-growth forest (Thomas et al. 1990). FEMAT (1993) evaluated numerous species in the range of the spotted owl with regard to old-growth habitat. They concluded "312 plants, 112 stocks of

Status and Functioning

TABLE 4-1. Summary of vital statistics of spotted owls from eleven intensively studied areas west of the Cascades, 1985-1993

	Mean	Standard error	Range
Probability of survival to age 1	.258	.036	.000-.418
Probability of adult survival	.844	.005	.821-.868[a]
Number of female young fledged per breeding female	.311	.125	.073-.524
Finite rate of increase	.925	.015	.830-1.019

[a]Includes second-year birds.
Source: Forsman et al. 1996

anadromous salmonids, 4 species of resident fish, and 90 terrestrial vertebrates were found to be closely associated with old-growth forest conditions." On the Olympic peninsula, spotted-owl home ranges are estimated at 1,680-2,800 ha (Lehmkuhl and Raphael 1993), with vertical canopy layering and large snag diameters being best predictors of owl presence in old-growth forest (Mills et al. 1993). The large sizes of home ranges suggest that habitat management strategies designed to maintain spotted-owl populations should also maintain habitat for other species dependent upon old-growth forest and the associated ecosystem processes within those forests. However, that owl populations are declining across the Westside range implies that sufficient habitat might not be available currently to perpetuate populations. Strategies for reversing that trend are to protect current old-growth forest used by owls and to identify and take advantage of opportunities to create suitable habitat in second-growth forest.

On BLM lands in western Oregon, spotted owl nest sites contained more old-growth forest and were in larger old-growth patches than randomly located sites in managed forests (Meyer et al. 1998). With the exception of old-growth patch size, no forest fragmentation indices were related to nest site selection. These forests consisted of approximately 25% old-growth and mature forest, as contrasted with estimates of approximately 60% of the Oregon Coast Range forests being in old growth before 1840 (Ripple 1994). Habitat occupancy may be related to environmental variables such as density of snags and down logs that are habitats for prey.

The situation is different in spotted-owl habitat in Eastside forests. Buchanan et al. (1995) reported that 60 of 83 nest sites examined in the Wenatchee, Wash., area were in forest in intermediate stages of succession, and the remaining 22 were in old-growth forests. Nesting was found in a wide range of stand ages, from 54 to 700 years. For 23% of the nest sites, the area had been partially logged 40 or more years ago. Specific characteristics of nest sites in this area were trees with 35-60 cm dbh with branches further above the ground than other trees within the area. Because food supplies and foraging efficiency were considered to have little influence on nest site selection, canopy height was considered important in increasing the ability of the owl to detect avian predators, such as the great horned owl. Management efforts to minimize fire in this area might have favored creation of suitable habitat in the short term but also might have increased the risk of habitat loss due to severe fire. In addition, fecundity rates of spotted owls on Eastside forests appear to be higher than on Westside forests. Over 5 years, Eastside rates averaged 0.49 female fledglings per female (range 0.11-0.74) (Irwin and Fleming 1991), compared with .31 female fledglings per female on the Westside (Forsman et al. 1996). The main conclusions from these comparisons is that the population dynamics and habitat relationships of spotted owls vary across the owl's range, that populations are declining west of the Cascade crest, and that insufficient habitat is now available on Westside forests to sustain owl populations.

The marbled murrelet is another threatened bird species of the Pacific Northwest. It is a seabird that nests in late successional forests along the coast and adjacent Coast range (Nelson et al. 1992). Gill-net fishing, oil pollution, and harvesting of old-growth forests within the range of the murrelet are considered threats. As part of an adaptive management approach, the Pacific Seabird Group (1993) recommended restriction of logging in known nest sites and similar habitats and restrictions in use of gill nets.

Bald eagles, listed as endangered in this region, are increasing after losses primarily attributable to pesticide contamination (Montana Bald Eagle Working Group 1991). However, pesticide residues in the lower Columbia River basin are still high and are affecting eagle productivity (Anthony et al. 1993). Nesting habitat in this region is primarily mixed-species, multistoried forest with snags and large trees or snags that protrude above the general forest canopy, within a mile of open water. On the Columbia River estuary of Oregon and Washington, bald eagles selected remnant stands of old-growth forest near shoreline for nesting habitat (Garrett et al. 1993). Because bald eagles are sensitive to disturbance, human disturbance needs to be minimized during nest-site occupation.

Flammulated owl, boreal owl, great gray owl (*Strix nebulosa*), white-headed woodpecker, black-backed woodpecker (*Picoides arcticus*), three-toed woodpecker (*Picoides tridactylus*), several hawks and owls, and many other bird species are dependent on old-growth interior forests. Pileated woodpeckers (*Dryocopus pileatus*) require forests with an old-growth component (McClelland 1979). Lewis' woodpecker is most abundant in forested communities with large trees (Robbins et al. 1983), which indicates that efforts to promote original gallery pine forests might benefit that species. Goldeneyes (*Bucephala clangula*), kestrels (*Falco sparverius*), Williamson's sapsucker (*Sphyrapicus thyroideus*), mountain chickadee (*Parus gambeli*), yellow-bellied sapsucker (*Sphyrapicus varius*), mountain bluebird, and common flicker (*Colaptes auratus*) all prefer to use large-diameter trees for nesting (McClelland et al. 1979).

Large mammals, including the grizzly bear, black bear (*Ursus americanus*), American marten (*Martes americana*), fisher (*Martes pennanti*), lynx (*Lynx canadensis*), and wolverine (*Gulo gulo*), inhabit forests in the Pacific Northwest and can be affected by management practices (Henjum et al. 1994; Ruggiero et al. 1994). Human-caused mortality is the major limiting factor for the grizzly across its range, and bears are at heightened risk in areas with roads. Logging from occupied grizzly habitat might improve forage conditions by creating herbaceous and shrubby habitat, but it also reduces habitat suitability by eliminating overstories used for security and resting areas. The intensified human activity that accompanies timber operations is an additional threat to the grizzly.

Other mammals, including the gray wolf (*Canis lupus*) and moose (*Alces alces*) rely on Pacific Northwest forest habitat. Restoration of the gray wolf to the northern Rockies presents management challenges similar to the grizzly, and timber-management guidelines that benefit the prey of wolf also should benefit the wolf and the species on which it preys. Moose habitat in central Idaho forests consists of closed-canopy forests, often with Pacific yew in the understory (Peek et al. 1987). Poaching is known to be an important limiting factor for moose in Idaho (Pierce et al. 1985).

The northern flying squirrel and Townsend's chipmunk are important prey species in Pacific Northwest forests (Carey et al. 1999). The flying squirrel consumes ectomycorrhizal fungi most abundant in late-successional forests. This squirrel is also a major prey species for the northern spotted owl. The chipmunk also consumes fungi, as well as seeds and fruits. These species, along with the red-backed vole, may be good indicators of functioning in these forests (Carey et al. 1999). Recommendations to conserve biodiversity have included thinning second-growth forests less than 50 years old at variable densities to increase crown differentiation, canopy stratification, and understory development. Retention of legacies by minimizing site preparation and burning so that ectomycorrhizal fungi link early in the successional process also have been recommended. Various activities to retain coarse woody debris, long, and snags in these forests could also be directed at maintaining fungi and associated small mammal prey species.

Amphibians and reptiles have begun to receive attention as concerns mount over their recent declines (Gibbons 1988; Raphael 1988). Older forests appear to support a greater diversity of species of herpetofauna than younger forests (Raphael 1988; Welsh and Lind 1988), but Raphael (1988) pointed out that structural characteristics, including coarse woody debris, hardwood understory, and abundance of moist sites are the critical criteria that dictate presence or absence of amphibians and reptiles. In addition, forest fragmentation might have an isolating effect on herpetofauna, which disperse across unsuitable habitat with difficulty. No Pacific Northwest forest amphibians are listed as threatened or endangered. In Idaho, the Coeur d'Alene salamander, which inhabits moist talus adjacent to forested areas, is classified as a species of special concern (Groves and Melquist 1990), and nine species of herpetofauna are listed for monitoring status in Washington.

Salmon and Other Fisheries

Anadromous salmon (salmon that spawn in freshwaters and migrate to the ocean) and steelhead trout in the Pacific Northwest are rapidly declining in abundance. Substantial numbers of native genetically distinct lines (stocks) have become extinct within the past century. Although debates over causes for the declines are heated, the status of many stocks is not in dispute. Most of the salmonid stocks of the mainstem Columbia River have been declining in abundance throughout the past century and have decreased sharply since 1970. Salmonids of coastal streams have exhibited similar, although less consistent, declines. Habitat and water-quality degradation, often associated with forestry and other land-use practices, are widespread regional problems. Few areas of high-quality salmon habitat remain, and many of the best remaining sites are in forested headwaters (NRC 1996).

Declines in Pacific salmon have been noted since the late 1800s. As the nation's attention turned to the West Coast and the California gold rush, logging and salmon fishing became major commercial industries in the Pacific Northwest. The first commercial salmon harvest occurred in the Columbia River in 1861. The first hatcheries were established on the McCloud River of California in 1872 and on the Clackamas River of Oregon in 1876 in response to decreases in salmon caused by overfishing and habitat degradation (Stone 1897). By the turn of the century, Washington, Oregon, Idaho, and California had passed laws to limit the freshwater harvest of salmon, and statutes were enacted to prevent private damming of streams and dumping of material into surface waters.

Resource-management agencies in the Pacific Northwest have raised concerns over the continuing decline of salmon throughout the twentieth century and developed numerous regulations to protect fish and habitats. Local areas or specific stocks of salmon have benefitted from these actions, but because of geographic scope and cumulative effects of human activities, Pacific salmon have continued to decline. Indeed, in response to the continuing problem, Oregon banned commercial and sport salmon fishing in 1994, but reopened it in 1999.

Recent analyses of available information on specific stocks of the five species of Pacific salmon document the regional and pervasive extent of the loss of salmon (NRC 1996). The Northwest Power Planning Council (1986) estimated that the numbers of salmon in the Columbia River basin declined from 10-16 million before the mid-nineteenth century to fewer than 2.5 million fish in the 1970s. In 1987, the U.S. Fish and Wildlife Service examined trends in salmon numbers for Alaska, Washington, Oregon, Idaho, and California during 1968-1984 (Konkel and McIntyre 1987). Thirteen of the 657 salmon populations for which adequate data were available became extinct during those 17 years. Significant trends, either increasing or decreasing, were observed for 30% of the populations—populations in Alaska tended to increase, but 24% of the populations in the Pacific Northwest declined significantly during the study period.

In 1991, the American Fisheries Society examined available evidence on the status of anadromous salmonid stocks in Washington, Oregon, Idaho, and California and concluded that more than 106 stocks have become extinct. Of the remaining stocks, 160 were classified as at serious risk of extinction, and an additional 54 were considered of special concern. Salmon of the Columbia River have undergone the greatest proportional loss of stocks, reflecting the history of intense commercial fishing, loss of habitat, and mortality associated with dams.

Subsequent evaluations of the status of salmon in the Pacific Northwest support the conclusions of the American Fisheries Society and identify additional stocks or habitats that have been lost or face serious risks. The Wilderness Society (1993) formed a panel of regional experts that concluded that salmon habitat on public and private lands has been seriously impaired by land-use practices, including forestry. All species of anadromous salmonids in the Pacific Northwest have undergone stock extinctions within the past century, and pink salmon is the only species in which the majority of stocks are not known to be declining.

Causes for the declines of anadromous salmonids are numerous and include habitat loss and conversion of streamside forest to agricultural, range, residential, or urban lands. Forest practices on existing forest lands, agricultural practices, grazing, dams that block passage, excessive commercial harvest, and sport harvest also contribute to declining populations (NRC 1996).

Over the past several years, scientists have established links between ocean conditions and upwelling, which brings nutrients into offshore environments. Production and survival of mature fish in the ocean is tied to climatic cycles in the North Pacific and to fishing pressure. Evidence of long-term cycles can be seen in the comparison of coho in Oregon and Washington and pink salmon in Alaskan fisheries. Catches of adult coho in Oregon and Washington are low when catches of pink salmon are high in Alaskan waters, changing over cycles of 20-35 years. But historic cycles are a small part of the current declines in salmon. The declines are related to offshore conditions as well as forest and other land-use practices.

Long-term records of cutthroat trout in the Alsea watershed in Oregon demonstrate that populations in watershed dominated by clearcuts are still at less than one-third of historic levels. The decrease is not attributable to ocean conditions, because cutthroat numbers in adjacent basins with mature forests are still at levels observed before harvest. Much of the population in the area is resident and not linked to the ocean environment.

Numerous other factors have contributed to the decline in salmon populations, including marine-mammal predation. Marine mammal protection has allowed populations of seals and sea lions to recover, and they consume returning adult salmon in the estuaries and mouths of coastal rivers. However, most analyses estimate that salmon make up a minor portion of the diet of marine mammals, and the additional mortality has a minor effect on salmon populations.

Invertebrates

Invertebrate species characterizing disturbed or early successional systems typically are adapted to wide variation in environmental conditions. Rather than being characteristic of particular communities, they often are present as parts of relatively nondistinct species assemblages. (That is, early successional assemblages tend to be made up of widespread, disturbance-adapted species that are not specific to a particular location. They are, of course, distinct from later successional assemblages, as demonstrated by Schowalter (1995)). For example, many of the herb and shrub species characteristic of early successional communities in Westside forests (e.g., snowbrush (*Ceanothus velutinus*), manzanita (*Arctostaphylos*), fireweed (*Epilobium angustifolium*), and daisies) are present in various early successional or frequently disturbed meadow and shrub communities from Alaska to northern California, and the species of aphids, ants, and other invertebrate species associated with those communities are widely distributed in western North America (Furniss and Carolin 1977).

Species closely associated with late-successional forests typically depend on the moderate conditions and resources provided by that forest structure. The organisms often cannot survive the more extreme conditions of open or disturbed areas (Seastedt 1984; Schowalter et al. 1986; Schowalter 1989). Many invertebrates, especially those that disperse by flying or ballooning, are capable of moving extensively in search of resources and can stray or be blown into unsuitable areas. But efforts to survive and reproduce do not necessarily contribute to stable species populations. For example, Schowalter (1989) found grass- and crop-feeding insects at the tops of old-growth Douglas-fir, and Edwards (1987) documented an extensive invertebrate deposition on glaciers on Mt. Ranier. Schowalter (1995) found old-growth Douglas-fir canopies had the highest invertebrate biodiversity, followed by mature Douglas-fir, shelterwood Douglas-fir, and then regenerating Douglas-fir, which had substantially lower invertebrate diversity. A largely different suite of invertebrate species lived in the canopies of old-growth hemlock (Schowalter 1995). Thus, the mix of Douglas-fir, hemlock, and other plants characteristic of old-growth, but not younger, forests substantially increases invertebrate biodiversity. Remnant mature trees in thinned stands (shelterwoods) can provide important refuges for many invertebrate species. Late-successional forest thus provides resources critical to survival of many species that, in turn, contribute to productivity of those forests. Many arboreal and forest-floor arthropods are abundant only in old-growth forests (e.g., Schowalter 1995; Winchester 1997), but most species remain poor known.

Fungi

Mutually beneficial associations between fungi and plant root systems are frequent in plant communities around the world (Harley and Smith 1983; Brundrett 1991). Fungi and plants are physiologically interdependent, with plants supplying the energy needs of fungi and fungi providing the plants with nutrients taken from the soil that plants need but are unable to access in sufficient quantities. Symbiotic fungi also can defend plants against root pathogens, and they protect plants against heavy-metal toxicity (Vogt et al. 1987; Vogt et al. 1991; Wilkins 1991).

Mushrooms, including those collected for human consumption, are the reproductive structure of filamentous fungi that form "ectomycorrhizal type" associations with trees. The Pacific Northwest is one of the prime areas for mushroom collecting in the world, in the dry Eastside

Ponderosa pine forests and the wet Westside Douglas-fir and hemlock forests (Molina et al. 1993; Pilz and Molina 1996).

In the dry and wet forests of the Pacific Northwest, mushrooms are an important food source for small mammals, and some species, such as the California red-backed vole (*Clethrionomys californicus*) , are almost totally dependent on mushrooms for their subsistence (Fogel and Trappe 1978; Maser et al. 1978). The California red-backed vole is found in young, mature, and old-growth forests in the Pacific Northwest but is common in old-growth (Maser et al. 1978). In western Oregon coniferous forests, some small mammals disappear from areas where trees have been harvested and only reappear when the regenerating forest reaches the pole-sapling stage, because they are dependent on the coniferous forest canopy (Ure and Maser 1982).

Mycorrhizal fungi are crucial to ecosystem processes because they facilitate and accelerate the rate at which plants are able to re-establish and recover after disturbances (Trappe and Luoma 1992). In Eastside and Westside forests, the maintenance of the symbiotic associations in forests is correlated closely with the presence of large, coarse, woody debris. Large, coarse wood is important in the ability of symbionts to survive over the short term (within annual cycles of drought) (Harvey et al. 1978) as well as over the long-term (e.g., from one disturbance to the next). The direct effects of the loss of mycorrhizal fungi or their propagules associated with forest-management practices on regeneration is dramatically shown in a study conducted by Perry et al. (1992). Forest-harvest practices (e.g., applying herbicides to deciduous shrubs that maintained the inoculi) caused loss of fungal inoculum, resulting in failure of woody revegetation at the site.

> *Land-use and forest-management practices have greatly influenced populations of numerous species, and have placed some of these species in danger of extinction.*

VIABLE POPULATIONS AND THE CONSERVATION OF BIODIVERSITY

Land-use and forest-management practices have greatly influenced populations of numerous species, and have placed some of these species in danger of extinction. To prevent extinction, viable populations must be managed. A viable population is the number of individuals "that will insure (at some acceptable level of risk) that a population will exist in a viable state for a given interval of time" (Gilpin and Soulé 1986). That can refer either to local populations or metapopulations, depending on the circumstances.[2]

The number of individuals that constitute a viable population is difficult to determine but depends in part on

- The population structure, social dynamic, and breeding characteristics of the species in question.
- Environmental fluctuations, particularly the possibility of catastrophic events that sharply reduce population size (Shaffer 1981).
- Environmental stresses, such as pollution, that reduce the vigor of individuals.
- Various aspects of habitat quality, including, in particular, the manner in which habitats are arrayed across the landscape — e.g., in large blocks, isolated fragments, or fragments connected by habitat bridges.
- The period for which viability is being assessed.

Populations that drop below some minimum size are drawn into what Gilpin and Soulé (1986) call the extinction vortex — i.e., they will become extinct unless extraordinary measures are taken. Populations that are losing individuals become at risk before that point, however, and when the population reaches a size near the extinction vortex, a random environmental event (such as drought, unusual weather, or disease) may draw the population into the vortex. Most information concerning minimum population size comes from models of vertebrate populations. They have not been tested for the plants, insects, and microbes that constitute the vast majority of Earth's biota.

Population size affects the survival chances of individuals in two general ways (Soulé 1983, 1986; Lande 1988): one related to demography

---

[2]In this discussion, "population" is used to cover both possibilities. For example, saving enough old-growth Douglas-fir to support 100 spotted owls will not save the owl, because it takes more than 100 individuals to maintain a viable population, and probably more than 1,000.

(the patterns of population growth and decline), the other to the ability retain genetic variability and avoid inbreeding.

Demographic Factors

Populations of most species fluctuate depending on numerous factors in their environment; the smaller a population is, the greater is the chance that one of its down cycles will either lead to its extinction or reduce it such that genetic or social deterioration triggers a slow slide to extinction. The population size at which a slide to extinction begins is partly determined by population genetics—especially inbreeding and loss of genetic diversity. Nongenetic factors also are important. In some species, small populations have relatively high genetic diversity, yet still are at risk because of random variation in population size.

The risk of extinction from demographic variations in population size is particularly high where factors such as infectious disease or large-scale natural events (e.g., hurricanes and wildfires) have the potential to reduce population sizes sharply (Hanson and Tuckwell 1981; Lande 1993). Small populations do not contain sufficient individuals to be buffered against such catastrophic losses. A particular threat is the possibility that two or more events that reduce population size occur in quick succession, not allowing the population time to recover from one before it is further reduced by another. Such multiple threats are most likely on landscapes where human activities are greatest.

Some animals depend on numbers for defense, foraging, or effective breeding. Such species have a social threshold population size below which the group becomes dysfunctional and unable to persist. In many cases, the social threshold is considerably higher than that determined solely by demographics or genetics (Soulé 1983).

Catastrophic events of one kind or another occur in almost any environment, and local populations of many species may disappear periodically from a given region. A species is not threatened by this as long as other healthy populations can provide a source of immigrants to replace local losses. However, when those other populations are in decline or highly fragmented and isolated, losses within local ecosystems may not be replaced, and extinction may occur. It is estimated that 100,000 to 300,000 species worldwide are threatened with extinction via this mechanism (Culotta 1994). The area required to buffer species against demographic calamities can be quite large: 200,000 ha or more for some species that live in forested habitats (Shugart and Seagle 1985).

Genetic Factors

Small populations sizes usually result in loss of genetic variation, particularly if the populations remain small for any length of time. Small, isolated populations can be at risk genetically for two reasons. First, fewer individuals exchanging fewer genes leads to inbreeding, which frequently results in loss of vigor and inability to resist stress (Lande 1988). Second, small populations tend to lose genetic variability through random processes called genetic drift.

An effective population size is the number of breeding adults that would provide the rate of inbreeding observed in a population if mating were random and the sexes were equal in number. Therefore, effective population size is the number of males and females that actually contribute equally to the gene pool, on average, from generation to generation In many species, many individuals do not breed or produce relatively few offspring; the effective population size is much less than the actual population size. For most mammals and birds, effective population sizes are no more than 25-50% of actual population size. Animals that live in family groups, such as wolves, typically have only one mating pair per group. Soulé and Wilcox (1980) estimated that an actual population size of 600 wolves would be required to maintain an effective population of 50 that prevented inbreeding depression.

The genetic factors involved in population sizes for plants have additional aspects not encountered in higher animals. Plants rely on some intermediary—wind or animals—for pollen and seed dispersal. Most pollen falls in the immediate neighborhood of the contributing parent, and plants with seeds dispersed by animals depend totally on the welfare of those animal vectors, including the area needed by the vectors to maintain viable populations. In addition, trees, probably because they are long-lived, accumulate more harmful genes than animals, and hence are likely to require a high level of outbreeding to prevent inbreeding depression (Ledig 1986). Management-related genetic selection for desirable growth traits has the potential to artificially narrow the gene pool of a whole species—selected and wild trees — because selected individuals presumably contribute to the gene pool in proportion to their numbers (Ledig 1986).

Population Viability Analysis

Theoretical considerations and empirical observations suggest that when genetic and demographic factors are accounted for, populations of animals and plants must contain at least several thousand individuals to be viable (Whitford 1983; Soulé 1987; Thomas 1990). However, no single number is applicable to all species, nor does a single number necessarily apply to any one species in all environmental situations (Gilpin and Soulé 1986; Thomas 1990).

Population viability analysis (PVA) can be used to identify threats faced by a species and evaluating the likelihood that it will persist for a given time into the future. PVA models take into account the view of habitats as landscape phenomena and can incorporate numerous features, including demographic stochastisity, environmental uncertainty, natural catastrophes, and genetic uncertainty. Conservation and recovery planning often must account for those and other variables, particularly because many endangered species exist in small populations, and without appropriate planning, a single event might destroy an entire species (NRC 1995; Akçakaya et al. 1999). Several PVAs have been conducted in recent years, including ones for the marbled murrelet, the northern spotted owl, and the red-cockaded woodpecker.

5
FOREST SUCCESSION, FIRE, AND LANDSCAPE DYNAMICS

INTRODUCTION

Disturbances such as fire, wind, and insect and pathogen outbreaks occur naturally in all Pacific Northwest forest types, although the frequency, intensity, and spatial extent of such events vary considerably. Natural disturbances—especially fire—shape the distribution and abundance of many forest species and influence key ecosystem processes across Pacific Northwest forest landscapes. Much of the debate and discussion regarding the status and management of Pacific Northwest forest ecosystems focuses on assertions regarding the status and dynamics of landscapes before European settlement and the manner and extent to which current management practices simulate them. Among those assertions are

- Suppression of fire and, to a lesser extent, other disturbance factors have resulted in ecosystem changes (e.g., the accumulation of fuels) that put landscapes at risk for catastrophic disturbance.
- In some cases, fire, insects, and pathogens disturbed forests with return times that roughly approximated timber rotations.
- Logging activities represent a reasonable surrogate for natural disturbances.

This chapter presents a conceptual framework for understanding the factors responsible for varying patterns of natural disturbance and the successional processes that derive from them. The specific role of fire with regard to the dynamics of Pacific Northwest forested landscapes is discussed, with particular attention to the effects human activities have on fire regimes across the region. Finally, the effects of disturbance associated with forest management and timber-related activities are compared with those of natural disturbance processes.

CONCEPTS OF SUCCESSION AND LANDSCAPE DYNAMICS

Many of the land- and natural-resource management policies and protocols developed in the first half of this century were strongly influenced by a comprehensive theory of succession described first by Cowles (1910) and subsequently refined by Clements (1916, 1928), both of whom worked primarily in forested regions. Central to that theory was the view that any disturbance initiated a linear and directional successional process culminating in a stable climax community with a structure determined largely by the constraints of climate. Such climax communities were thought to be composed of species able to reproduce successfully in the context of small-scale disturbances, such as tree-fall gaps. Large-scale disturbances, such as fire, were viewed largely as the result of human interventions, and presettlement North America was thought to have been dominated by vast, unbroken expanses of climax forests. Climax communities were thought to be the most stable species assemblages, and it seemed to follow that successional change proceeded on an inexorable path toward increasing stability (Odum 1969).

Over the past 3 decades, virtually every aspect of this classical theory of succession has come under question (e.g., Drury and Nisbet 1973; Connell and Slatyer 1977; Egler 1977; Sousa 1984), and the change in understanding of succession has major implications for the management of Pacific Northwest forests. We now recognize that fire and other disturbance processes operate over a wide range of spatial scales and were indeed an integral part of the primeval landscapes. Through evolutionary time, recurrent disturbances of varying intensity and frequency have selected for adaptations in the biota, making some species more resistant to and other species more resilient from the effects of disturbances. Life histories of many tree and other species have evolved to be dependent on such disturbances.

Successional change does not necessarily lead to increased resistance to disturbance. Rather, successional change might facilitate future disturbance. For example, as forests develop, changes in structure, species composition, and the accumulation of organic matter alter fire susceptibility. Those changes may increase the likelihood of fire or other disturbances, as happens with the ingrowth of shade-tolerant firs and accumulation of woody fuels in Eastside ponderosa pine forests. Conversely, old-growth Douglas-fir and ponderosa pine forests tend to be less susceptible to crown fires than younger forests (Perry 1988a; Franklin et al. 1989). They are well buffered against wind; hence, they remain relatively moist. Tree crowns are high in the air, so the probability of fire moving from the ground into the crowns is low, at least in the absence of understory fire "ladders" (smaller understory trees). The thick bark of old-growth Douglas-fir and ponderosa pine also increases their likelihood of surviving a ground fire, and ground fires (which historically were frequent in many forest types) reduce fuel loads while leaving the overstory of older trees frequently scarred but still intact.

Disturbances vary in intensity, spatial extent, frequency, seasonality, and predictability. That variation results in different patterns of response often involving different suites of species within a given ecosystem (Agee 1993). A landscape can be viewed as a collection of disturbance-reduced patches of varying size and undergoing change that is conditioned not only by the history of disturbance of a patch, but also by the character and position of the patches that surround it, against a background of yearly and decadal variation (Delcourt and Delcourt 1991). The importance of the overall forest matrix on the patterns of change within individual stands or patches makes a landscape approach to understanding and managing ecosystems compelling.

The temporal analog to spatial relationships—"legacies" (Franklin 1993b)—provides additional complexity. Disturbances do not wipe the slate clean, but rather leave legacies such as residual debris, seed banks, advanced regeneration, and surviving plants and animals that influence future trajectories of change. The nature of such legacies depends on the status of the ecosystem before the disturbance (e.g., evenly aged thinning forest versus mixed-age old-growth forest), as well as the behavior (intensity and variability) of the disturbance itself. The legacies vary among fire regimes and between natural disturbances (such as fire and wind storms) and management interventions (such as logging).

Current theories of succession, although less simplistic than the classical ones, do postulate the existence of predictable patterns of change following catastrophic disturbances. One description of the pattern is a four-stage model of forest succession (see Peet and Christensen 1987):

- The first stage — *establishment* — is characterized by extensive seedling establishment and rapid growth, during which competitive interactions among individual plants are somewhat limited.
- Closure of the tree canopy initiates the second stage — *thinning* — during which competition among more or less evenly aged, individual trees is intense. Overall stand biomass continues to increase rapidly during this stage, but the number of trees decreases as shaded and less-vigorous individuals die. Shade-tolerant trees may gradually invade the subcanopy during this stage.
- Eventually, tree density (number of trees per unit area of land) decreases to the point where individual tree deaths create gaps that cannot be filled by the lateral growth of adjacent trees. That initiates the third stage — *transition* — during which new individuals begin to fill gaps, creating an unevenly aged forest. During this stage, living biomass might remain constant or decrease while total soil and above-ground biomass might increase, owing to the accumulation of woody debris.
- In the final stage — *steady state* — the first three stages occur as a mosaic of patches within the forest (Watt 1947; Bormann and Likens 1979). How frequently this final stage occurs in most forested landscapes is uncertain (Peet and Christensen 1987).

Variations on this four-stage process are common. For example, in some areas of the Pacific Northwest, the earliest forest canopy may be formed as an even-aged stand of red alder that is subsequently replaced by a more or less even-aged stand of Douglas-fir. Where the red alder cover is particularly dense, seedling establishment of Douglas-fir may be inhibited. Where seedling establishment is sparse and initial tree density or stocking generally low (e.g., many ponderosa pine forests), competitive thinning (stage 2) might not occur.

Small-scale disturbances in forests (e.g., gaps created by the loss of one to a few trees) tend to facilitate the replacement of Douglas-fir by shade-tolerant species. An old-canopy, Douglas-fir tree that dies as part of natural tree mortality is likely to be replaced by hemlock or cedar seedlings expanding into the canopy opening (Spies et al. 1990). Similarly, winter storms that blow down isolated trees or groups of trees usually do not open sufficient areas to permit Douglas-fir regeneration. If sufficient area is opened to permit regeneration, root rots can prevent seedling survival. Consequently, small-scale disturbances typically promote regeneration of shade-tolerant hemlock and red cedar on the Westside. However, if windthrow provided sufficient fuel accumulation, a subsequent fire could denude the site and reinitiate succession. In the absence of fire that eliminates hemlock and cedar, a Douglas-fir forest eventually will be replaced by a more diverse forest of shade-tolerant species. In other situations, patterns of fuel accumulation and frequency of ignition are such that the process is interrupted before reaching the later stages. Shade-tolerant trees may not become successfully established in forests with frequent ground fires (as in ponderosa pine and Douglas-fir on drier sites).

FIRE AND LANDSCAPE DYNAMICS

Among the various sources of natural disturbance, fire has been most important in affecting the forested landscapes of the Pacific Northwest. The frequency and intensity of fire within the region varies considerably depending on temperature and moisture conditions of the site, ignition patterns, and the characteristics of individual tree species and their susceptibility to fire (Agee 1981). The most obvious differences occur between the west and east sides of the Cascade Mountains. Fires are less frequent in the moist climate of the Westside, and fire-return intervals are relatively long and highly variable. More frequent drought on the Eastside, coupled with different patterns of fuel accumulation, favor shorter fire-return intervals with relatively low-intensity surface fires (Agee 1993).

The relationship between moisture availability and fire frequency is complex (Martin 1982). Biomass production and the rate at which flammable fuels accumulate particularly are correlated with available

Forest Succession, Fire, and Landscape Dynamics

water. In seasonally dry areas, such as the Eastside, fuel production is sufficiently high to carry fires at more frequent intervals, and dry conditions that are favorable for ignition and fire spread are likely. Increased moisture and humidity favor higher rates of fuel production; however, drought periods during which fuels dry sufficiently to burn decrease in frequency.

Among the various sources of natural disturbance, fire has been most important in affecting the forested landscapes of the Pacific Northwest.

In general, fire intensity (expressed as fuel consumed or energy expended per unit area) is inversely related to frequency. In forests typified by short fire-return intervals, the flammable fuels of the forest floor are spatially separated from the flammable needles and branches of the canopy. Fires in those forests tend to burn only litter and small woody debris. Where fires burn less frequently, fuels are distributed more continuously from forest floor to canopy. Fires in these forests are more likely to spread to tree crowns.

Presettlement Fire Regimes and Successional Change

Before 1850, infrequent, severe fires ("stand-replacement fires") with a highly variable return interval of more than 100 years were common in the western hemlock/Douglas-fir forests, Pacific silver fir forests, and subalpine forests (Agee 1993). Stand-replacement fires are followed by major changes in plant-species composition (Agee 1993). Moderate-severity fires—with a 25- to 100-year interval between fires—were common in dry Douglas-fir, mixed evergreen, red fir, and lodgepole pine forests (Agee 1993). Moist Douglas-fir forests in the central Cascades had a moderate fire regime, at least during the 1800s (Morrison and Swanson 1990; Teensma et al 1991). Fires of moderate severity often resulted in stands with two or more age classes of trees (Agee 1993). Low-severity fires—with a 1- to 25-year interval between fires—were typical in oak woodlands, Ponderosa pine, and mixed-conifer forests and tended to maintain those ecosystems, which are tolerant of fire (Agee 1981).

Huff (1984) noted that variations in fire regimes are highly correlated with the relative prevalence of Douglas-fir and hemlock. Douglas-fir, although relatively tolerant of fire, is intolerant of shading, whereas hemlock and cedar are less tolerant of fire but tolerant of shade. When fire-return time is 100-300 years, Douglas-fir is dominant; longer return times result in a mixture of Douglas-fir and ponderosa pine. Fire-return times exceeding 600 years favor hemlock dominance. The long intervals between fires on the Westside favor Douglas-fir domination early in the successional sequence, but late-successional forests increasingly become dominated by western hemlock; hemlock begins to replace Douglas-fir at about 250 years. It can take several hundred more years without fire for western hemlock to attain stand dominance (Agee 1993). A considerable portion of the old growth of Westside forests was characterized by long-return fires (interval of more than 500 years) (Huff 1984).

In areas dominated by spruce and hemlock, very moist conditions and rapid decomposition of fine fuels result in fire return times of 500-1,100 years (Fahnestock and Agee 1983; Huff 1984). In northern California and southern Oregon, where the coastal redwood is important, fire regimes are much more variable. Return times of 400-500 years can be the norm in moist areas, and more frequent surface fires (every 50 years) typify drier locations. Severe winds can disturb these forests more frequently (Ruth and Harris 1979; Harcombe 1986), and large-scale (many hectares in extent) blowdowns occur in these forests with a return time of 300-400 years (Harcombe 1986).

On the Eastside and over much of the Idaho-Montana region, frequent low- and moderate-intensity fires maintain an open, parklike forest dominated by fire-tolerant species, such as ponderosa pine and larch. Those forests often have open canopies with a heterogeneous understory of grasses and shrubs. Low-intensity fires have been frequent, with return intervals of 4-7 years common in some places (Arno 1980; Bork 1985). Frequent fires have kept the stands open, providing sites for advanced regeneration of pines as canopy trees die from windthrow, disease, or occasional fire "hot spots" (Gruell 1985; Merrill et al. 1980).

Succession after low-intensity fire events is rapid. Large trees are generally unaffected, and most herbs and shrubs resprout rapidly from belowground rootstocks, bulbs, and burls. Intense disturbances, such

Forest Succession, Fire, and Landscape Dynamics

as crown-killing fires, favor invasion by shrubs such as greenleaf manzanita and deer brush. The mineral seedbed and open light conditions after low-intensity fires favor heterogeneous patterns of pine establishment and thinning in localized patches. With subsequent thinning, pine canopies are kept open by repeated low-intensity surface fires. Long fire-return intervals in these forests favor growth of the shrubby understory species and invasion of shade-tolerant firs and Douglas-fir.

On the Eastside and in Idaho and Montana, lodgepole pine forests are typical of higher and moister elevations. Surface fires occur as frequently as every 50-80 years and stand-replacement fires occur with a 150-300 year return time. The intervals are considerably longer and less predictable on the moist west slope of the Cascades. In those areas, lodgepole stands often are unevenly aged, and disturbance factors other than fire, such as bark beetles, can be far more important in determining stand structure.

Large forest fires in 1910, 1919, and 1934 burned much of northern Idaho and western Montana, creating a variety of successional stages from shrub field to young conifer stands, which characterize much of the area at present (Wellner 1970; Arno 1980; Gruell 1985). Interspersed among the regrowth are old western hemlock, western red cedar, and Douglas-fir. Many of those communities are mixed-species forests, with western larch and western white pine also prominent. In the more southerly portions of the region, a history of frequent fires has also contributed to the development of a mosaic pattern of forests. Extensive stands of lodgepole pine characterize higher-elevation forests, while a complex of Douglas-fir and ponderosa pine forests dominate at lower elevations.

Montana and Idaho forests dominated by Douglas-fir generally have longer fire intervals than ponderosa pine forests. Grand fir, western red cedar, and western hemlock forests have still longer intervals, but when burned (or logged and burned), a tall shrub complex develops that persists for as long as 50 years and that is eventually replaced as the conifer forest develops (Mueggler 1965; Wittinger et al. 1977; Crane et al. 1983; Stickney 1986; Morgan and Neuenschwander 1988). Higher-elevation spruce-fir forests might be replaced by lodgepole pine in the overstory when burned, but understories characteristic of the spruce-fir community resprout and are not replaced by a tall shrub complex (Lyon 1984).

Human Alteration of Fire Regimes

Over the past century, human activities in the Pacific Northwest have altered fire regimes enormously. On the Westside, development of urban areas and transportation networks, increased recreational use of forests, and timber activities have increased the frequency of fires in many areas (Pyne 1995). Residual fuels and patterns of post-fire succession have facilitated successive and sometimes devastating reburning of many areas (e.g., the Tilamook fires) (Pyne 1982). But regionwide generalizations are difficult because human land use and transportation corridors have broken the landscape and altered the movement of fires in complex ways that are not totally understood.

Foresters have long recognized the differences in susceptibility to fire of different age classes of forests. Andrews and Cowlin (1940), in their analysis of wildfires in western Oregon and Washington between 1924 and 1933, noted that, "A much higher proportion of the Douglas-fir seedling and sapling areas, of recently cut-over land, of old deforested burns, and of the noncommercial types is burned over annually than of saw timber areas." Because flammable fuels are more continuously distributed vertically and available for burning in early successional forests than they are in old-growth or late-successional forests, early successional forests are more susceptible to intense fires than their more mature counterparts. The extent and landscape-level continuity of those forests has increased owing to fires, timber activities, and other human actions, resulting in more flammable conditions in many areas of the Westside.

Over the past century, human activities in the Pacific Northwest have altered fire regimes enormously.

Fire hazard has increased throughout much of western Oregon and Washington because of the increase in area of young forest plantations (Perry 1988a; Franklin et al. 1989). It has been known for many years that plantations are more vulnerable to crown fires than are healthy old-growth forests (e.g., Andrews and Cowlin 1940; Cowlin et al. 1942).

Forest Succession, Fire, and Landscape Dynamics

Once a landscape reaches some critical proportion of susceptible types of stands, disturbances propagate across boundaries and affect even relatively unsusceptible types (Perry 1988a; Turner et al. 1994).

Human activities have undoubtedly increased the frequency of forest ignition on the Eastside as well; however, active suppression of fire has played a much greater role in determining the current status and fire regimes of forests in this region. Fire suppression and selective logging of old-growth trees on the Eastside has led to encroachment of fire-intolerant and pest-susceptible Douglas-fir and true firs in those forests (Anderson et al. 1987; Wickman et al. 1992; Mutch et al. 1993; Covington et al. 1994). That in turn has led to more devastating fires and pest outbreaks in recent years than likely occurred in the past when fuels and host abundance were limiting factors (Anderson et al. 1987; Agee 1993; Hessburg et al. 1993; Wickman et al. 1992; USFS/BLM 1994).

Wildfire-control programs over the past 75 years have been enormously successful in reducing the area burned each year, nationally and regionally (MacCleery 1995). At the same time, those programs have created "a different fire regime, one characterized by uncontrollable fires burning in heavy fuels" (Mutch et al. 1993). Increased vulnerability to crown fires in Eastside forests can be traced to the same factors that have exacerbated insect and drought problems (see Chapter 3). The combination of successional ingrowth associated with fire exclusion and the development of more densely stocked stands after logging has increased the likelihood of extensive surface fires in the region. That condition probably extends to higher-elevation forests as well (Perry 1988a).

In northern Idaho, where severe fires are more frequent than in most of Oregon and Washington, tree mortality after fire was approximately 70% in mature stands, 80% in thinning stands, and more than 90% in early-successional stands (Hutchison and Winters 1942). Some of the high mortality in young stands on cutover sites in the early part of the century might have been due to burning residual logging slash. However, even with more thorough slash removal, young stands are more vulnerable than mature forests and old-growth forests. For example, in the 1987 Silver Fire—the most severe in southwest Oregon in at least 50 years—45% of the old-growth stands had less than 10% mortality, but this was true of only 18% of the small sawtimber stands (USFS 1988a).

NATURAL DISTURBANCE AND HUMAN MANAGEMENT: AN ECOLOGICAL COMPARISON

Much current debate regarding forest-management practices centers on the relationship and similarities between the patterns of natural disturbance and change discussed above and silvicultural practices (Hansen et al. 1991, Swanson and Franklin 1992; Perry 1995a; Kohm and Franklin 1997). Patterns of ecosystem response (e.g., nutrient transformations, changes in species composition, and accumulation of flammable fuels) vary considerably among fire and silvicultural regimes as a function of predisturbance history, site conditions, disturbance characteristics, and postdisturbance environment. Furthermore, it is not necessarily the case that simply because a disturbance is natural (having happened without human intervention), it is more benign or more conducive to sustaining forest health and ecosystem processes than forest-management activities (Christensen 1988; Christensen et al. 1989).

Landscape Considerations

Patterns of natural disturbance, especially fire, were highly variable on landscapes before European settlement. Landscape pattern influences where fires burn and their intensities. When weather conditions are extreme (as they were during the Yellowstone fires of 1988 or the great fires of 1910 in Idaho) fires may cross major patch boundaries, and a new mosaic is created (Turner et al. 1994). Without intervention, fires will burn up to natural boundaries, such as rivers; ridges; and low-fuel ecosystems, such as rocky scree or feldfields. Except in the most extreme situations, variable fire behavior results in environmental variability, even within burn boundaries.

The spatial scales over which timber and other resource management activities typically take place rarely are those for fire, and the boundary conditions usually are quite different. Boundaries of logging units often are related to ownership boundaries typically do not coincide with ecosystem boundaries, such as watersheds and ridges. The checkerboard patterns of cutting on railroad lands represents one extreme example of such arbitrary patterns. Harvest activities, especially clearcutting, generally leave a much less variable postdisturbance environment than do fires. It is possible to devise silvicultural systems that mimic patterns typical of natural disturbances (Hansen et al. 1991; Swanson and Franklin 1992; Perry 1995b). However, patterns of ownership and economic considerations can make such systems impractical, and more research is needed to understand the potential benefits of such systems with regard to management goals (see Chapter 8).

Human development and building on fire-prone landscapes has increased the financial liability and risk to human life associated with catastrophic fires that can occur as a consequence of extensive fuel accumulations. Furthermore, development severely constrains the variety of management interventions that can be used to remedy the situation. For example, prescribed fire in heavy fuels close to homes or other structures is risky and expensive and can be applied on only limited scales. The use of prescribed fire as a management strategy is limited in some places owing to the effects of smoke on air quality near already polluted urban centers.

Fuels

Logging is often cited as a means of reducing fuels and, thereby reducing fire danger, particularly in areas where fire suppression has resulted in considerable fuel accumulation and where public liability from fire is high. But the direct effect of fire on fuels is highly variable. Repeated, low-intensity fires tend to produce fuel loads and structures (e.g., vertical stratification) that are not conducive to high-intensity, crown-killing fires. However, crown-killing fires often result in an increase in the load of flammable fuels in the form of charred trunks, and some of the largest burn complexes in the Pacific Northwest (e.g., the Tillamook Burn) occurred as reburns.

Logging has been proposed as a possible surrogate for fire in reducing fuel accumulations with the added benefit of economic return (Agee 1993), but logging or clearcutting do not necessarily reduce flammable fuels. Residual slash can carry intense fires, and rapid regeneration of early-successional shrubs and trees can create highly flammable fuel conditions within a few years of cutting. Without adequate treatment of small woody residues, logging may exacerbate fire risk rather than lower it (Agee 1993). If cutting such as that allowed under recent salvage logging legislation is to reduce fuel accumulation, considerably greater attention must be paid to logging and postharvest management practices that will accomplish that goal.

Nutrient Fluxes

Fire results in immediate oxidation of large amounts of carbon and rapid fluxes of nearly all nutrients. The loss of nutrient capital (especially nitrogen and phosphorus) can be substantial and is largely dependent on fire intensity and the quantity of fuel consumed. In fine fuels (small-diameter fuels) burning at high temperatures, as much as 70% of the nitrogen in the consumed fuel can be volatilized or gasified (phosphorus loss is usually considerably lower); however, in heavy fuels, that loss is usually more in the range of 20-40%. Nutrients not volatilized are added to the soil as char and ash, and a considerable portion of the above-ground nutrient capital can remain in burnt snags and woody debris that decompose slowly.

Fire often results in an immediate flush of nutrient availability as a consequence of ash and a postfire environment conducive to mineralization (e.g., Grier 1975; Christensen 1985). That nutrient flush is important to the establishment of many forest plant species, and the removal of the litter mat and exposure of mineral soil by fire is important to successful germination and establishment of many conifer species. Increased nutrient mobility also increases nutrient loss to leaching, adding to the overall loss of nutrient capital, although those losses for nitrogen amount to less than 1% of the total prefire capital. The flush of nutrient availability is short-lived (a year or two), and in some cases, subsequent soil-nutrient availability can be lower than prefire conditions.

It is conventional wisdom that the loss of nutrient capital owing to forest cutting is considerably greater than that due to fire, and it seems obvious that clearcutting coupled with slash burning results in large nutrient losses. However, carefully controlled studies comparing nutrient budgets among different fire regimes or cutting treatments have not been conducted.

Whether nutrient losses associated with natural or human-caused disturbance are detrimental in the long term depends on the patterns and rate of recovery of nutrients. In the case of phosphorus or cations, replenishment of nutrient capital is primarily by atmospheric inputs and can be quite slow. Nitrogen fixation accelerates inputs of nitrogen. Although claims have been made that fire increases activities of nonsymbiotic, nitrogen-fixing microbes, solid evidence is lacking. But the importance of symbiotic, nitrogen-fixing plants, such as buckbrush and red alder, in postfire successional ecosystems is well known.

Biological Diversity

Given the close tie between the life history of some species and fire (e.g., serotinous cones in lodgepole pine and heat-stimulated germination in some shrubs) the responses of many species to fire differ from their responses to cutting. The greatest differences are found in the years immediately after disturbance. The high degree of spatial variability generated by burning probably facilitates higher levels of species diversity than less-variable postharvest environments, such as those typical of clearcuts (Christensen 1988). Where management is intensive, many silvicultural activities are explicitly designed to diminish diversity and increase production and yield of target species.

6
PRODUCTS FROM THE FORESTS

The forests of the Pacific Northwest are the source of wood products—including lumber, plywood, pulp, and paper—and nonwood products—such as fish and wildlife, recreation, and assorted miscellaneous products, including ornamental greens and foods. All of those products have or could have monetary value in ordinary markets. The forests also provide scenery, clean water, and clean air, amenities that are less readily bought and sold but that are valuable nonetheless. This chapter focuses on wood products and selected nonwood products. It examines the implications of changes in the use and management of Pacific Northwest forests for those products.

The charge to the committee in its statement of task was drafted before adoption of the Northwest Forest Plan and at a time when the region's wood products economy and management of its forests was changing rapidly. Substantial changes have since taken place as a result of that plan, but other changes have also occurred as controversies continue over the management and future uses of the region's forests. One result of these changes has been a sharp decrease in timber harvests on federal land in the Pacific Northwest, as well as in much of the rest of the western states. As implied in the statement of task, this has brought about shifts timber harvests in other regions that also supply national markets for wood products. This section of the report describes these changes in timber harvests and some of their implications for sustaining other forest values.

PACIFIC NORTHWEST WOOD PRODUCTS IN THE NATIONAL ECONOMY

The Pacific Northwest wood products economy is driven by the United States market for softwood lumber and structural panels (softwood plywood and oriented-strand board (OSB), a panel product made of wood that competes with softwood plywood). New home construction and home repairs and improvements are the major factors in this market, which has been expanding in recent years despite the cutbacks in federal timber harvests.

National consumption of softwood lumber increased by 10% from peak to peak of the wood products cycle from 1977 to 1997 (Figure 6-1). National consumption of all softwood wood products, excluding fuelwood, also increased over this period (Figure 6-2). Worldwide consumption of wood for industrial purposes has also increased in recent years and is expected to continue to increase over the next few decades (Brooks et al. 1996). At the same time, per capita consumption in the United States of the kind of timber that is the main product of Pacific Northwest forests—softwood sawlogs and softwood veneer logs—shows no strong upward trend (USFS 1994).

Increasing demand for wood products nationally has been met over the years through various market responses. For example, softwood timber supplies have been extended by improving access to timber, increasing the number of tree species that is used, using wood more efficiently, substituting plentiful hardwoods for scarcer softwoods, using products made of wood particles in place of solid wood products, gluing small pieces of lumber together to make larger pieces, and recycling. Higher prices for wood products have also led to some use of substitute materials, such as aluminum and vinyl siding. Some opportunities for increasing the efficiency of wood use remain, such as using modular units in construction; most of these opportunities are at the processing stage. Greater efficiency in wood use, including conservation measures, likely will continue, driven for the most part by increases in wood prices.

The region west of the Great Plains, including the Pacific coast states and the intermountain and Rocky Mountain regions, was the nation's major softwood-lumber-producing region, but its share of the nation's total softwood lumber production has fallen from more than 70% in the

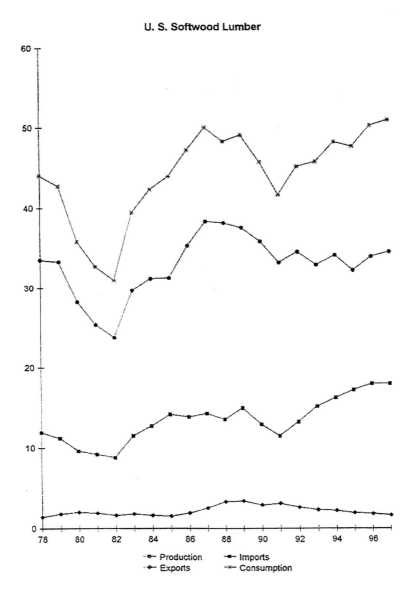

FIGURE 6-1. U.S. production, imports, exports, and consumption of softwood lumber 1978-1997 (in billion board feet). Source: Howard 1999.

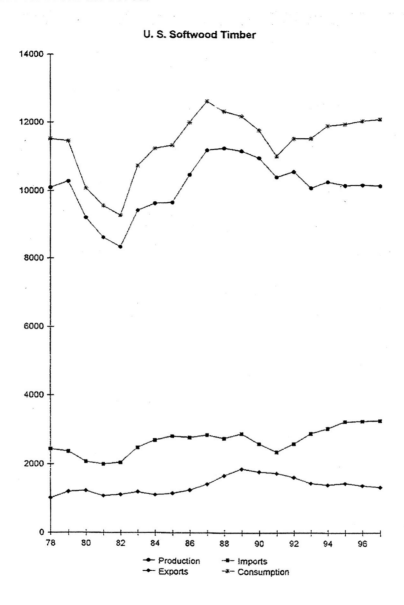

FIGURE 6-2. U.S. production, imports, exports, and consumption of softwood timber products, excluding fuelwood, 1978-1997 (in million cubic feet roundwood equivalent). Source: Howard 1999.

mid-1960s to about 55% now (Adams et al. 1988). Washington and Oregon were the biggest lumber producers for many years, but their share of the nation's total has fallen from more than 40% in the mid-1960s to about 30% recently (Adams et al. 1988).

The role of federal lands in supplying the timber that is processed into softwood lumber and plywood has changed. From 1962 to 1989 in Oregon, timber harvests from federal lands were generally higher than those from private lands, but harvests on federal lands are now well below private harvests (Figure 6-3). While timber harvests from privately owned forests have exceeded those from federal forests for many years in Washington, the spread has widened considerably since 1988 (Figure 6-4).

Douglas-fir and hemlock from the region west of the Cascades summit are the main softwood lumber species from the Pacific Northwest that compete generally in construction markets with softwood lumber from British Columbia, the Rocky Mountains, the southern United States, and eastern Canada. Sitka spruce and cedars go into specialty markets that depend on the characteristics of the particular species. Hardwood lumber production, mostly red alder, has increased in the Pacific Northwest in recent years, but it is still a minor factor in the region. Ponderosa pine and western white pine from the interior of the Pacific Northwest (eastern Oregon and Washington, Idaho, and Montana) are used mainly for millwork (doors, windows, and molding) that goes into national markets. Other softwood species from the entire region (spruce, lodgepole pine, and fir) are used to make lumber that goes into national markets for construction materials.

Little softwood lumber in the region is processed beyond the sawmill or planing mill before it is used in construction. Market factors determine that most use of softwood lumber in manufacturing takes place closer to the point of final use. Fiber used in production of wood pulp and paper in the Pacific Northwest depends heavily on chipped residues from lumber and plywood mills. That source is being supplemented increasingly with recycled paper and with hardwood pulpwood. Some chips from lumber and plywood mills are also exported, mainly to Japan.

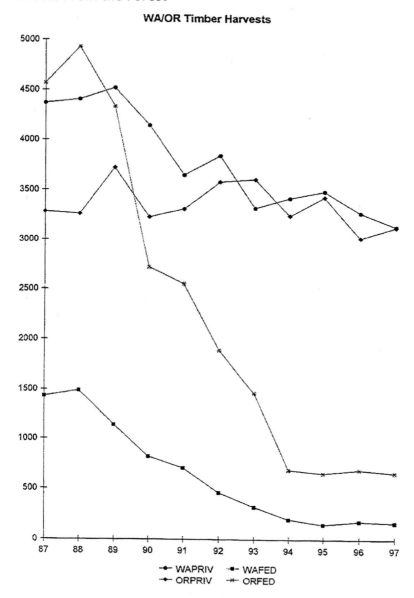

FIGURE 6-3. Federal and private timber harvests by state, Washington and Oregon, 1987-1997 (in million board feet). Source: Warren 1999.

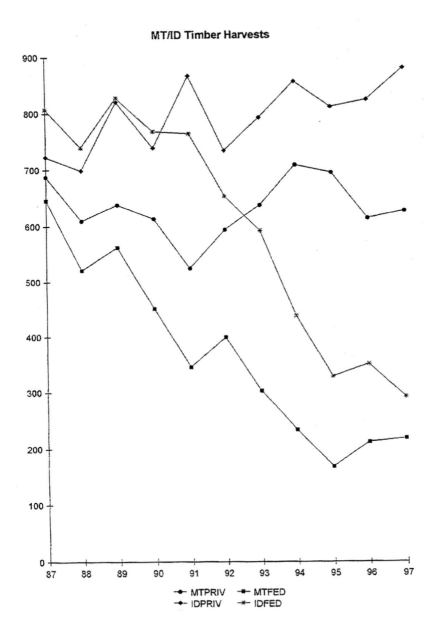

FIGURE 6-4. Federal and private timber harvests by state, Montana and Idaho, 1987-1997 (million board feet). Source: Warren 1999.

PRODUCTS IN NATIONAL AND INTERNATIONAL MARKETS

The Pacific Northwest was the major source of softwood lumber and plywood for U.S. markets for many years, but production in the southern U.S. now exceeds that in the Pacific Northwest by significant margins (Figure 6-5 and Figure 6-6). In addition, imports from Canada now account for about one-third of U.S. softwood lumber consumption, about the same as the southern United States (Figure 6-7).

The Pacific Northwest also once produced nearly all of the nation's softwood plywood, mainly from large old Douglas-fir. The South started producing pine plywood in the early 1960s and became the leading producer of softwood plywood by 1980 as plywood technology improved and markets changed to accept plywood with knots for structural uses. As timber prices increased in the late 1970s, first Minnesota, Wisconsin, and Michigan and then the Northeast and South began producing panels made of wood particles, such as oriented-strand board (OSB), which competes directly with softwood plywood in construction. National OSB production now exceeds that of softwood plywood from the Pacific Northwest, which has fallen from its peak levels in the 1960s and 1970s (Figure 6-6).

> *The Pacific Northwest was the major source of softwood lumber and plywood for U.S. markets for many years, but production in the southern U.S. now exceeds that in the Pacific Northwest by significant margins.*

Softwood log exports are now less than one-half of what they were in the late 1980s. The volume of log exports (about 2 billion board feet) is equivalent to about one-quarter of the Pacific Northwest softwood lumber production. Softwood lumber exports have grown to the point where they are now close to the same volume as log exports (Howard 1999).

EFFECTS OF CHANGES IN FEDERAL TIMBER HARVESTS IN THE PACIFIC NORTHWEST

Any substantial change in Pacific Northwest timber harvests leads to changes in the markets for wood products, as well as to changes in other

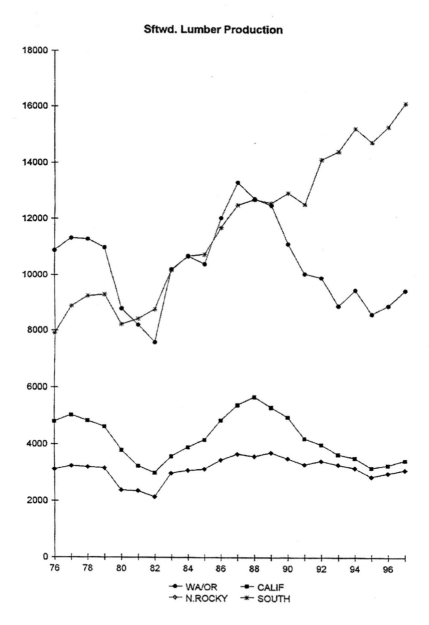

FIGURE 6-5. Softwood lumber production by U.S. region (in million board feet). Source: Adams et al. 1988.

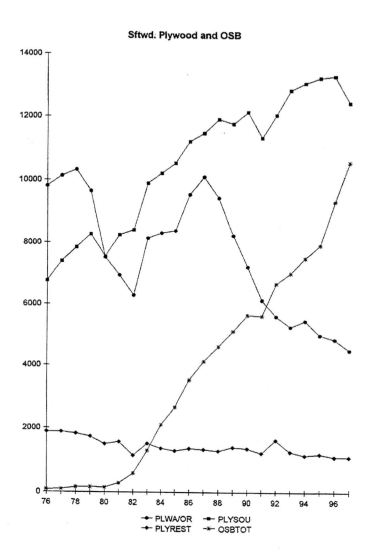

FIGURE 6-6. Annual softwood plywood and oriented strandboard (OSB) production by U.S. region, 1976-1997 (in million square feet, 3/8" basis). PLYWA/OR = softwood plywood production in Washington and Oregon; PLYSOU = softwood plywood production in the South; PLYREST = softwood plywood production in the rest of the U.S.; OSBTOT = U.S. oriented strandboard production. Source: Adams et al. 1988.

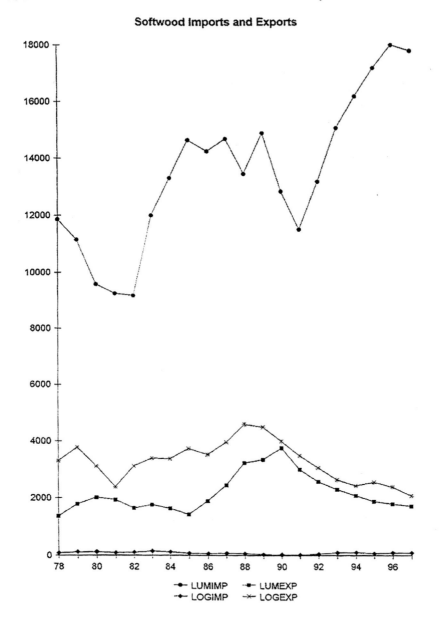

FIGURE 6-7. U.S. softwood lumber and softwood log imports and exports, 1978-1997 (in million board feet). Source: Howard 1999.

related forest outputs. In a market economy, changes in timber harvests are reflected in shifts in economic supply and demand. Such shifts lead to price changes and to income and substitution effects (Hicks 1946). Income effects occur because price changes affect consumers' buying power — giving them more if prices fall or less if prices rise. Substitution effects occur because of changes in the relative prices of competing goods.

Shifts in markets for wood products have been substantial as a result of the cutbacks in federal timber harvests over the past decade. The nature and extent of the shifts were not, however, wholly unexpected. Even though these markets are complex, economic models used to simulate the operation of the market economy provided useful insights. Two such economic models of the U.S. timber economy — the Timber Assessment Market Model (TAMM) and the CINTRAFOR Global Trade Model (CGTM) - helped policy makers assess the likely effects of changes. TAMM was developed by the U.S. Forest Service (USFS) in cooperation with Oregon State University and is used by the USFS in much of its policy analysis (Adams and Haynes 1980). CGTM was developed at the University of Washington (Cardellichio et al. 1988). Actual changes in markets since implementation of the Northwest Forest Plan were reasonably close to those indicated in advance by analyses made using these models once producers and consumers had time to react and adjust manufacturing processes and consumption patterns.

Changes in the markets for wood products occasioned by the cutback in Pacific Northwest federal timber harvests spread throughout the nation's timber economy (Haynes and Adams 1992). Prices for standing timber (stumpage) rose throughout the West during the late 1980s and early 1990s as federal timber sales were reduced. The sharpest increases occurred west of the Cascades in Washington and Oregon as the President's Northwest Forest Plan was implemented. But timber prices also fell again, especially in western Washington and Oregon, as the timber industry began adjusting to the new situation. At the same time, prices for softwood timber in the U.S. South continued to go up, although their peak in the late 1990s was about the same as their peak in the late 1970s (Figure 6-8). This occurred as the effects of reduced timber supplies from federal forests in the West spread to other regions, but they soon spread to other regions whose wood products replaced

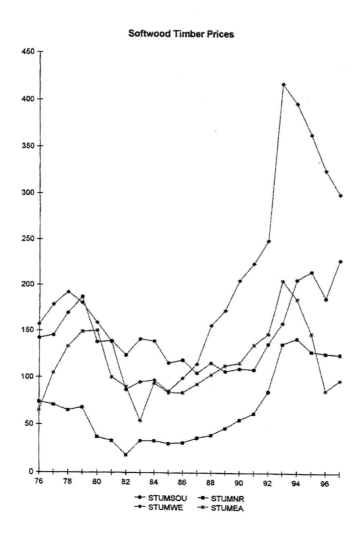

FIGURE 6-8. Softwood sawtimber (stumpage) prices, deflated (1982 prices by U.S. region, 1976-1997 (per 1,000 board feet). STUMSOU = South; STUMNR = Northern Rockies; STUMWE = western Washington and Oregon; STUMEA = eastern Washington and Oregon. Source: Adams et al. 1988.

Products From the Forest   *135*

those that would have otherwise come from the Pacific Northwest (Haynes and Weigand 1997). Five general kinds of responses are important and relevant:

- Increased harvests at the extensive margin (forests that previously were uneconomic for harvesting)
- Increased harvests at the intensive margin (forests that already were supplying wood to markets)
  - Increased use of hardwoods
  - Technology changes
  - Materials substitutions

Increased Harvests at the Extensive Margin

General price increases for timber will make some previously submarginal timber economic for harvesting. That is the kind of effect that historically led to the lumber industry moving from region to region as timber resources in each area were depleted. The major areas of unexploited softwood forests in North America at the time the Northwest Forest Plan was adopted were in eastern Canada and, to a lesser extent, in western Canada. The United States has practically no softwood timber (other than that in reserved status) that is not already under some level of management and is not already considered part of the market supply.

Siberia has a large volume of potentially marketable softwood timber that is not now part of the world's market supply. Imports of either softwood logs or lumber from Siberia or other regions, other than those from Canada, have been negligible (Howard 1999). Imports of unprocessed or untreated softwoods pose a significant problem in the possible introduction of harmful nonindigenous species (Office of Technology Assessment 1993a). This can have serious impacts on biodiversity in general and on forests in particular (NRC 1999b). The United States now prohibits imports of various unprocessed wood products, but questions have been raised about the adequacy of these controls in preventing importing pests.

The substantial increases in softwood lumber imports from Canada since adoption of the Northwest Forest Plan raise questions about their sustainability. Forestry Canada, the national government forestry agency, indicated in 1993 that timber harvests had reached allowable cut limits in some areas and that British Columbia might reduce its harvests, largely in response to pressure from environmental interests (Forestry Canada 1993). Relying on harvesting primary forests in other regions to substitute for Pacific Northwest harvests may provide only short-term relief in the absence of substantial investments in more intensive forest practices.

Increased Harvests at the Intensive Margin

Some substitution for reduced federal timber harvests in the Pacific Northwest has come from forests, especially private forests in the South and the Pacific Northwest, that will have to be managed more intensively (the intensive margin) to sustain the additional harvests. Much of these private forests are already managed for timber production. Sustaining these higher harvest levels will require added investments in forest management.

The most recent available USFS assessment of the forest resources situation (USFS 1994) indicates that softwood timber removals (harvesting and land clearing) in the South exceeds growth by 14%, a reversal of the long-standing situation of growth exceeding removal (USFS 1994). Although the 1994 assessment notes that situation might again change in another couple of decades, some analysts suggest caution. Cubbage and associates, for example, suggest that "sustaining long-term timber inventory and harvest increases in the South will be extremely difficult in the areas where the wood is needed the most [areas of concentrated wood industry plants]" (Cubbage et al. 1995).

The USFS has identified several practices that could be used to increase timber growth in the South. The top ones in terms of their contribution to increased timber growth are planting with site preparation, clearcutting mature stands and regenerating stands, eliminating trees that compete for growing space with crop trees, and commercial thinning of overly dense stands of merchantable trees (USFS 1988a). One indication of the level of investment is tree planting on privately owned land. A steady upward trend since 1950 in the area planted each year peaked at about 3 million acres annually in the late 1980s. It has been nearly level at about 2.5 million acres in the 1990s (Moulton et al. 1993; Moulton 1998). But these totals mask to a degree the effect of more intensive timber management practices on forest industry land relative to that on other private land (Moulton 1998). Forest industry lands, which make up only 14 percent of the nation's timberland, account for over one-third of the timber harvest, a proportion that is likely to go up because of increasingly intensive management (NRC 1998).

The role of intensively managed tree plantations in South America, New Zealand, Australia, and Africa in providing wood products is also likely to increase and, to a degree, substitute for federal timber from the Pacific Northwest in world markets. Because of high growth rates relative to forests in temperate zones, one estimate is that plantations, often of nonindigenous species such as eucalyptus, that are managed intensively could supply the equivalent of the current world demand for industrial wood from as little as 4 percent of the global forest area (Sedjo and Botkin 1997). So far, however, the direct role of products from such plantations in U.S. markets has been minimal. For example, changes in U.S. production, imports, and exports of paper and paperboard (the products of tropical forests most likely to compete effectively in U.S. markets) show that while U.S. production went up by 16.4 million tons and exports went up by 5.1 million tons from 1990 to 1997, imports went up by only 2.5 million tons (Howard 1999).

Increased Use of Hardwoods

Increased use of hardwoods in place of softwoods has been taking place for years in response to increasing prices for softwood timber. Cutbacks in federal timber harvests will speed this substitution. Increased use of hardwoods is likely in the north-central and northeastern United States, where more aspen will be used to make products that compete with softwood plywood and with softwood lumber used for millwork, and in the South (Haynes et al. 1995).

Potential shortages in the West of chips for making wood pulp are leading some firms to plant fast-growing, irrigated poplars as a supplementary source of wood fiber. Natural stands of hardwoods suitable for making pulp are generally thought to be in plentiful supply in most of the eastern states, but planting fast-growing species, such as poplar, might become economical in the East as pressure on natural hardwood forests mounts.

Technology Changes

Price increases for timber lead to higher costs for users and are a spur to innovation. In the past, development of plywood lowered labor costs in construction. Then, the development of OSB made it possible to use low-cost timber as a substitute for the higher-cost softwood timber used in making plywood. Parallel developments, such as laminated veneer lumber and recycling (mainly for fiber products such as paper), are likely to be a further response to higher softwood timber prices. Other technological developments in sawmills and plywood plants driven by high timber prices have made it possible to use smaller logs in making lumber and plywood and for getting more lumber or plywood out of the same volume of timber (Office of Technology Assessment 1993b; Haynes et al. 1995).

Materials Substitution

Human needs, such as shelter and furnishings, being met by wood products can also be met using other materials. The choice of wood is usually influenced by price, as well as by more subjective preferences, such as experience and culture. Metals, plastics, and concrete appear to be increasingly used as substitutes for wood. In particular, steel studs used in framing houses and aluminum and vinyl siding are substituting for Pacific Northwest wood products. Recycled wastepaper is also used in making fiber-based products to an increasing extent in the United States and worldwide (Ince 1994).

With the 1993-1994 increase in softwood lumber prices, steel studs became nearly price competitive with softwood studs. If the price of wood studs becomes greater than that for steel studs, tradition and familiarity with wood are unlikely to be sufficient barriers to prevent a significant shift in use in favor of steel. The use of recycled wastepaper in papermaking has been, in good part, driven by policies. Cost advantages, apart from those imposed by changes in public policies, do not appear to have played a big part in this increase (Haynes et al. 1995).

ENVIRONMENTAL EFFECTS

The reduction in federal timber harvests affects environmental values. The positive environmental effects on federal lands in the Pacific Northwest are described in FEMAT and the Interior Columbia Basin Ecosystem Management Assessment (USFS 1996). The potential negative effects occur mostly on nonfederal forest land in the U. S. and Canada where timber harvests have increased in response to the reductions on federal lands in the U. S. The U.S. South and Canada account for most of the substitution for reduced federal softwood timber harvests. Although some observers expected increased harvests from private forests in the Pacific Northwest in response to the reductions in federal harvests, private harvests have remained more or less level since 1987 in Oregon, Montana, and Idaho and have fallen somewhat in Washington (Figures 6-3 and 6-4).

Increased timber harvests on private land in the U.S. South and on provincial land in Canada have responded nearly equally in substituting for the reductions in Pacific Northwest harvests from federal land (Figures 6-5 and 6-6). These regions are the focus of the discussion of environmental effects below. The increased harvests on private lands involve some environmental costs, as well as some economic costs (USFS 1990). Substitution of other materials for wood products in the marketplace, an alternative to increased timber harvests, would also entail some environmental costs (Koch 1992).

In Canada, the increased timber harvests since adoption of the Northwest Forest Plan have come from forests that were economically submarginal prior adoption of the plan. These forests typically are less productive and have less economically usable timber per acre than private forests in the Pacific Northwest and the U.S. South. The extent of the environmental effects associated with these increases in harvests is related to the additional area that is harvested and to the specific character of the forests in which the harvesting occurs.

Increased production of timber on private forests in the South has meant lower average ages of trees at harvest and more intensive silvicultural operations in response to higher timber prices. These have meant conflicts with environmental values such as protection of wetlands and habitat for endangered species such as the red-cockaded woodpecker.

State laws, as well as federal regulations, guide and constrain forest management practices on private forests. The five Pacific Northwest states have some of the strongest laws regulating forest practices in the country (Ellefson et al. 1995). The extent to which more intensive forest management will increase environmental effects on private lands depends on the varied conditions under which such management occurs and the effectiveness, extent, and enforcement of state regulations. Increasing forest growth in the South to make future increases in timber harvests possible will require significant increases in forest-management investments. That might require public incentives such as cost-share payments and technical assistance to landowners in addition to the incentive of higher timber prices (NRC 1998).

Extending softwood timber supplies by substituting hardwoods for softwoods in some uses is an ongoing process that may have been speeded up by the reductions in federal timber harvests. Most of the potential expansion in hardwood harvests is in the eastern United States, where concerns have been raised about the effects on sustaining productivity, impacts on wetlands, and effects on endangered species.

The likelihood of more intensive forest management in other regions of the United States as a result of cutbacks in federal timber harvests in the Pacific Northwest raises questions about the ability of the nation as a whole to pursue the goals of forest management identified in this report.

The other way of dealing with reduced softwood timber supplies is to use substitutes for lumber and plywood, such as metals, plastics, and concrete. These materials require more energy in production and use than do wood products, which may result in greater environmental impacts than producing and using the wood products that they replace (NRC 1976). Other environmental effects associated with the use of these substitute materials could include effects on surface resources,

Products From the Forest

water quality, and air quality. Estimates of energy use or the effects on other resources depend on the specific conditions of production and use of the substitutes.

IMPLICATIONS FOR OTHER REGIONS

The likelihood of more intensive forest management in other regions of the United States as a result of cutbacks in federal timber harvests in the Pacific Northwest raises questions about the ability of the nation as a whole to pursue the goals of forest management identified in this report. For example, can viable populations of indigenous forest species be sustained in the eastern United States if wood products production increases to fill the gap caused by the reduction in Pacific Northwest harvests? Can ecological processes in general be sustained (especially in the South) as increasing pressure is put on its forests to produce softwood lumber and wood pulp?

Several important characteristics of eastern forests distinguish them from those of the Pacific Northwest. The eastern United States has almost no untouched old-growth forest and has more hardwood forests than softwood or coniferous forests (NRC 1998). Forest ownership is mainly private, mostly owned by nonindustrial owners, although federal and state-owned forests are important in some areas (Table 6-1 and Table 6-2).

TABLE 6-1. Area of Forest Land (in Thousand Acres) by Major Class and Subregion in the Eastern United States, 1992

	Total productive			
	Forest land	Timberland	Reserves	Other
Northeast	85,380	79,449	4,550	1,382
North central	83,108	78,350	2,992	1,766
Southeast	88,078	84,794	1,998	1,285
South central	123,760	114,515	1,050	8,193
Total	380,326	357,108	10,591	12,627

Source: Powell et al. 1993.

TABLE 6-2. Timberland Ownership(in Thousand Acres) in the Eastern United States by Subregion, 1992

	National forest	Other public	Forest industry	Nonindustrial private
Northeast	2,179	6,498	11,858	58,914
North central	7,366	14,264	4,340	52,380
Southeast	4,847	4,309	16,252	59,387
South central	6,707	4,639	22,774	80,395
Total	21,099	29,709	55,223	251,076

Source: Powell et al. 1993.

The most intensive forest-management practices in the East have focused on softwood forests, especially in the South, but also to a degree in the northern coniferous forests from Maine to Minnesota. Softwood forests in the northeast and north central regions been under additional pressure as a result of reductions in federal timber harvests in the West. But most of the pressure has fallen on the South and on southern pine timber, which is managed for both sawlogs and pulpwood.

Private forests account for about 90% of the total timberland in the South. Of that, 20% is now owned by forest-industry firms, and 70% is owned by nonindustrial private owners. The forest-industry ownership is skewed toward the pines (longleaf-slash pine, loblolly-shortleaf pine, and oak-pine) and is mainly in the coastal plain and lower piedmont portions of the South (Powell et al. 1993). Much of this forest grew up on abandoned farm fields two or three generations ago; the farms had replaced pine forests previously maintained as fire-dependent types.

Industry-owned forests in the South provided 60% as much timber for the market as nonindustrial private forests did in 1984, although the area of industry forests is less than one-third that of nonindustrial forests. The share coming from industry forests is expected to be nearly equal to that from nonindustrial forests by 2010 because of rapid increases in the intensity of timber management on industry forests (USFS 1988a).

Intensive management for southern pine timber production on forest-industry land now typically consists of site preparation after logging by spraying herbicides and using controlled burning to control brush, hand planting genetically improved pine seedlings, fertilizing sites that have a specific nutrient deficiency, and clearcutting the trees 22-35 years after planting. Plantations are typically thinned to yield commercial products, mainly pulp wood, beginning at 15-20 years if the expected rotation age is greater than 25 years.

The intensity of management on nonindustrial private forests in the South varies widely. The USFS estimated that only 5% of the nonindustrial pine plantation acreage received the highest level of management (similar to that described above), while 25% of the forest-industry plantations received that level in 1980-1985 and 50% received it by 1985 (USFS 1988a). One key to increased softwood timber supplies from the South is the extent of pine plantations. In 1984, pine plantations provided 13% of the softwood timber that was supplied to the market. That was estimated to increase to 25% in 1990, to 43% in 2000, and to 58% in 2010 (USFS 1988a). Further increases in response to reductions in federal softwood timber harvests in the West are likely to come from an expanded area of plantings.

As in other regions, neither ownership patterns nor forest conditions are uniform over large areas. Forests owned by a firm tend to be an agglomeration of small (200-2,000 acres each) and often noncontiguous tracts generally within a radius of 25 miles from a plant. The tracts are intermingled with those of other industrial firms and nonindustrial owners. Most tracts have 10-15% or more of their area in hardwoods in strips along streams and in scattered patches.

About 65% of the total acreage planted in the United States in 1993 was in the nine southern states from North Carolina around the coast to Texas. The area planted in each state in 1993 was greater than 100,000 acres. The total area planted that year was down by about 120,000 acres from 1992 and by 975,000 acres (29%) from 1988, the peak year (USFS 1994). But the potential for increasing softwood timber production through planting is still high. The USFS estimated that 22 million acres of marginal cropland and pasture in the South would yield greater returns to the owners if planted to pine and that this would add about 2.1 billion cubic feet to annual softwood growth (USFS 1990).

Increasing timber production in the South has raised concerns about susceptibility of pine plantations to insect and disease attacks, effects on wetlands, and effects on species viability. All of these concerns have some merit, but to date, none has substantially limited increases in overall timber growth and production. Major disease threats have to some degree been countered with genetic modifications in pine plantation stock. The southern pine beetle has not had a major effect on overall timber production, because infested trees usually are harvested (Bechtold et al. 1992). Forest managers have avoided major effects on wetlands. Reduction in timber production to protect the major endangered species — the red-cockaded woodpecker — has been limited largely to areas with mature timber and has had little effect on the bulk of the private forests. If any of these factors changed, the ability of the South to respond to reductions in federal timber harvests in the West would be affected.

If investments in forest management and softwood timber production in the South are limited for some reason, the major burden of responding to reductions in timber harvests in the West will shift to other sources. These will be increasing use of alternative timber species, especially hardwoods, in the eastern United States and possibly to even greater dependence for wood on Canada and the southern hemisphere. The increased use of aspen and other soft-textured hardwoods to make oriented-strandboard, a substitute for softwood plywood in construction, is an example of the first of these responses. Inasmuch as eastern hardwoods have generally been in plentiful supply, the impacts of this increased use has been limited so far, and management of these species has not been greatly intensified.

Expansion of timber harvests since 1990 in boreal forests in Canada, especially in eastern Canada, has raised questions about the ability of these forests to sustain high harvest levels. For example, while the annual volume of logs harvested in Quebec went up by 94 percent from 1990 to 1995, the area planted and the area of site preparation each fell by 31 percent (Canadian Council of Forest Ministers 1997).

EFFECTS ON REGIONAL AND NATIONAL INCOME

The reductions in federal timber harvests in the Pacific Northwest have led to some shifts in income among regions and have created some winners and losers. Regions that have seen increases in timber harvests (i.e., the South) have gained while those that bear the brunt of the cutbacks are the biggest losers, as expected (USFS 1990; Adams et al. 1996). Total losses in regional and national income were expected to exceed total gains (Perez-Garcia 1993; Haynes and Adams 1992). Regional wood products producers are affected most, while the reductions in federal timber harvests have had little impact on consumers (Haynes and Weigand 1997). Possible negative impacts on workers and other dependent on federal timber harvests in the Pacific Northwest have been ameliorated by federal assistance provided by the Northwest Economic Adjustment Initiative that was part of the Northwest Forest Plan (Tuchmann et al. 1996).

CHANGE AND INCENTIVES

No change in Pacific Northwest forest management will be permanent. The situation is dynamic, just as it has been since the national forests were established at the beginning of the century. Markets for forest products will continue to change in response to various pressures, and our understanding of forest biology and management practices will continue to improve.

The most important forces driving change lie outside the forest sector. World population and income continue to grow. Public policy in the United States is committed to sustaining economic growth. The country's high rate of consumption and associated low rate of savings help to explain increases in consumption of all types of forest products, including wood products. Lifestyles based on low gasoline costs also put substantial pressure on forests for recreation and home sites, as well as for wood products, especially in building and remodeling suburban homes.

Pressures on forests for all uses in the Pacific Northwest and elsewhere in the United States will probably continue to rise. As overall demands on forest resources rise, the changes in use of federal forests in response to the Northwest Forest Plan and other initiatives will continue to change the range of incentives, market and otherwise, facing owners and managers of private forests in both the Pacific Northwest and other affected regions. The basic demands for materials, space, and environmental amenities will almost certainly continue to increase for the foreseeable future. For example, the emergence of markets for carbon sequestration, a new phenomenon brought about by growing concerns over global climate change, may create new opportunities and pressures for Pacific Northwest forests. These opportunities and pressures, which have come to the fore partly as a result of the Kyoto Treaty on Climate

Change, were not evident a decade ago. They could alter investment opportunities, especially those involving retention of mature forests, but also those for young, rapidly growing forests.

As overall demands rise, the changes in use and management of federal lands in response to the Northwest Forest Plan have changed the incentives facing owners and managers of other forests. Possible results of the changed incentives that operate through markets for wood products were discussed above. But the changes that will occur in the character of the Pacific Northwest forests relative to what is likely to have occurred in the absence of the plan, such as the added area of old-growth reserves, may lead to shifts in patterns of recreation and other aspects of lifestyles. Such shifts in turn are likely to create new and unknown incentives for management and use of private forests.

NONWOOD PRODUCTS FROM FORESTS

Dividing this chapter into separate sections on wood products and nonwood products inevitably will lead to comparisons of the treatment afforded each category. For the most part, wood products are bought and sold in ordinary markets that establish prices and values. The products are reasonably well defined and consistent and information on markets is collected and made widely available. The information can be used to evaluate effects on discrete forest areas. Nonwood products, on the other hand, are heterogeneous and often sold in poorly formed markets. Price and value information, as well as other information on markets, is often poor if it is collected and available at all. The information that is available often cannot be readily applied to discrete forest areas or situations.

The decisions that led to the reductions in federal timber harvests in the Pacific Northwest affect the output of both nonwood and wood products. Nonwood products include wildlife, fish used commercially and for recreation, outdoor recreation not tied specifically to fish and wildlife, water, amenities such as scenic landscapes, and a wide variety of minor forest products, including berries, ornamental greens, and mushrooms. In many cases production of these nonwood products competes in the forest with production of wood products. The lack of roughly equivalent kinds and quantities of information for each category impedes analyses of trade-offs that might help in making choices, such as those in the Northwest Forest Plan. At the same time, the wide range in the kinds of value systems involved in decisions about publicly owned forests and resources suggests the difficulty of finding a single calculus that fairly treats wood and nonwood products (NRC 1998).

Available information is used in the sections below to respond to the committee's charge in the statement of task that it "review the use of forest products from the Pacific Northwest and the degree to which forest products from other parts of the United States can be substituted for them." Most of the nonwood products of Pacific Northwest forests are not sold in national markets and do not have readily obvious substitutes from other regions of the United States. As a result, the available information does not permit a discussion of nonwood products that is parallel to that for wood products in the earlier sections of this chapter.

The following discussion presents some information on wildlife-related recreation, other forest-related recreation, fisheries, mushrooms, and water. The information is sketchy, but it allows for some discussion of the implications of the reduction in federal timber harvests in the Pacific Northwest, and the implications of the value of nonwood products for allocating forest land to wood and nonwood products.

Wildlife-Related Recreation

More than 75% of the people in the Pacific Northwest (excluding the part of California in the region covered by this report) participate annually in wildlife-related recreation, with at least 50% engaging in nonconsumptive appreciation of this resource (Table 6-3). More than one in ten people in the region hunt, and about three of ten people fish. Those proportions are higher than for the United States as a whole (USDOI 1993). Hunting, fishing, and nonconsumptive wildlife appreciation are not mutually exclusive—each might be engaged in concurrently. Declines from 1985 to 1996 in the proportions of the population engaging in wildlife-related recreation (USDOI 1993, 1998) are related to a decline in economic conditions during the earlier parts of the period and possibly to increased urbanization, both of which tend

TABLE 6-3. Participation in wildlife-associated recreation in the Pacific Northwest[1]

Recreation	Year	Idaho[2]		Montana		Oregon		Washington		Total[3]		U.S. Total[4]	
		n	%	n	%	n	%	n	%	n	%	n	%
Fishing	1985	287	41	225	37	689	34	1,057	32	2,258	34	46,357	26
	1991	247	33	171	28	540	24	967	26	1,925	26	35,578	19
	1996	275	31	158	24	501	20	834	20	1,768	21	32,222	16
Hunting	1985	169	24	153	25	306	15	296	9	924	14	16,684	9
	1991	161	22	158	26	240	11	251	7	810	11	14,063	7
	1996	183	21	141	21	272	11	25	6	852	10	13,321	4
Nonconsumptive	1985	645	91	545	93	1,728	86	2,886	87	5,804	87	134,697	74
	1991	385	52	312	52	1,124	51	2,076	56	3,897	54	76,111	40
	1996	855	40	315	47	1,048	42	1,621	39	3,339	41	62,868	31
Total	1985	656	93	556	92	1,743	87	2,921	88	5,876	88	140,078	77
	1991	578	77	469	78	1,615	73	2,919	79	5,581	77	108,745	57
	1996	484	55	394	59	1,260	51	1,908	45	4,046	49	76,964	38

[1] These are estimated participants by state of residence. Extracted from National Survey of Fishing, Hunting, and Wildlife-Associated Recreation (USDOI 1989, 1993, 1998).
[2] Percent of total state population 16 years and older.
[3] Percent of total for northwestern states.
[4] Percent for entire United States, population 16 years and older.

to reduce participation (Walsh et al. 1989). The least urbanized of the four states, Idaho and Montana, generally have higher participation rates than Oregon and Washington. Participation in big-game hunting increased nationwide from 1980 to 1996, although hunting for small game decreased. Hunting for migratory birds increased between 1991 and 1996, likely in response to increased waterfowl populations (USDOI 1993, 1998).

Flather et al. (1999) project a decline in big-game hunting in the Pacific Coast states and an increase in the Rocky Mountain states through 2050. Changes in participation in big-game hunting are related to amount of congestion on public hunting grounds, changes in land-use patterns, and family and work obligations. Satisfaction from big-game hunting derives from a variety of attributes in addition to success in taking game, including skills in woodsmanship and marksmanship, contact with nature, escape from daily routine, and companionship (Potter et al. 1973).

Three regions of Idaho with different ground conditions provide examples of different conditions for hunting elk and their implications for wildlife management (Table 6-4). Idaho's southeastern region, for example, provides hunters more than 5 times the opportunity to see elk than the more densely forested region of northern Idaho does. The elk numbers in southeastern, northern, and central Idaho are high (which indicate adequate habitat), but to prevent severely reducing the elk population, the hunting seasons must be shorter in the the open forest and rangeland habitat of the southeastern region. As a result, restricted-entry hunting, which limits the number of hunters in an area, is most common in that region.

The data for the three regions of Idaho suggest that forest-management decisions require close coordination between land management and the people who use the land. In particular, experience elsewhere in the United States suggests that participation and interest in hunting declines as hunting opportunities are constrained. Attempts in Oregon to address problems of low life expectancy for bull elk (which is attributable to heavy hunting pressure) by restricting hunting have led to a decline in the number of hunters and intense controversy. Efforts in Idaho to address forest-health issues by changing forest conditions from dense stands to more open stands, as has been suggested (e.g., O'Laughlin et al. 1993), will increase access and affect hunting conditions.

TABLE 6-4. Comparison of Hunting Statistics from Three Regions of Idaho, Illustrating the Effect of Access and Forest Conditions[1]

Region	Panhandle[2]	Backcountry[3]	Southeast[4]
Estimated elk population	9563	22369	13935
Elk (number/square mile)	1.88	1.86	1.51
Mean harvest of elk	242	282	124
Mean harvest square mile	0.048	0.024	0.013
Hunting season length (days)	14-24	38	9
Mean number of hunters	1500	1147	623
Mean percent success	18.1	23.9	14.7
Mean days afield per hunter	6.9	7.3	3.8
Mean animals seen	1.1	2.6	6.3

[1]Population estimates for units, 2, 4A, 5, and 62A projected from Toweill and Hanna 1985. Elk management plan, 1986-1990. Idaho Dept. Fish & Game, Boise. Other information for 1989-1993 period from Kuck, ed. 1992. Statewide surveys and inventory. Elk. Project W170R16, Study I, Job 1; Kuck, ed. 1993. Statewide surveys and inventory. Elk. Project W170R17. Study I, Job 1.; Unsworth et al. 1991. Elk management plan, 1991-1995. Idaho Dept. Fish & Game. Boise, 62pp.
[2]Northern Idaho, heavily forested habitat with moderate to high access, hunting season October 10-24 for antlered elk, and October 15-24 for antlerless elk in 1993. Hunting units 2, 3, 4, 4A, 5, 6, 7, and 9.
[3]Central Idaho, mountainous habitat with poor access and more open forests than northern Idaho. Hunting seasons in wilderness Sept. 15-Nov. 18, elsewhere October 10 - November 8, in 1993. Hunting units 10, 12, 16!, 17, 19, 19a, 20, 25, 26, 27, 34, 35, and 36. Bulls only during regular season. Permitted hunts for antlerless not included.
[4]Southeastern Idaho, open forests and rangeland which is highly accessible. Hunting seasons October 10-October 19 in 1993, bulls only during regular season. Permitted hunts for antlerless not included. Hunting units 51, 58, 59, 59A, 60, 61, 62, 62A, 64, 65, 66, 67, 69, 75, 77, 78.
[5]Harvest, number of hunters, % success, days afield, and animals seen are means for all units within region per year. Season length in days.

Products From the Forest

Expenditures by participants for wildlife-related recreation in the Pacific Northwest increased from 1985 to 1996 (Table 6-5). Differences among the states are apparent, with lower 1991 fishing and hunting expenditures than in 1985 in Idaho, Oregon, and Washington, and higher expenditures in Montana over the same period. Expenditures on wildlife-related recreation in the region, including those for trips and equipment, nearly doubled from 1985 to 1991, reaching $3.6 billion in 1991. The increase in the region (35%) was greater than that nationwide (6%). The increase in total expenditures came during a period of decline in wildlife-related recreation (down 5% in the region and down 22% nationwide), which indicates increased expenditures per capita.

Other Forest-Related Recreation

As the demand for outdoor recreation opportunities continues to increase, demand for particular kinds of opportunities changes in response to shifts in population, lifestyles, and interests. The changing trends in recreation use present no clear picture for the Pacific Northwest. Nationally, people are traveling shorter distances and spending less money per recreation visit, and developed-area camping is increasing while backcountry camping is decreasing (USFS 1988b). Physical activities, such as skiing, canoeing, and kayaking are gaining in popularity, and activities that present risks and adventure for the participant are expected to become even more popular (USFS 1988b).

Categorizing current recreation trips involving land, which includes forests but also much more, is problematical because of the wide range of definable uses. The largest number of trips nationwide in 1987 involved sightseeing (329 million), walking (273 million), pleasure driving (233 million), and picnicking (213 million). The least number of trips involved backpacking (13 million), visiting prehistoric sites (16 million), horseback riding (25 million), and primitive camping (camping in backcountry areas) (28 million). There are presumably significant regional differences in the relative rankings of these kinds of recreation use that are related to the opportunities that are afforded for them. Rates of projected increase do not appear to be vastly different for the various uses mentioned above. Data on recreational use of wilderness areas on national forests are intriguing. Total use increased from 1971

TABLE 6-5. Expenditures, in Thousands of Dollars, for Wildlife-Associated Recreation in the Pacific Northwest

Recreation	Year	Idaho	Montana	Oregon	Washington	Total	U.S. Total
Fishing	1985	106,791	81,618	226,559	322,295	737,363	13,280,738
	1991	74,829	121,376	204,744	284,635	685,578	11,847,750
	1996	279,950	243,501	622,806	704,396	1,850,653	37,797,061
Hunting	1985	68,168	45,082	79,535	84,426	277,251	3,714,194
	1991	44,245	105,698	56,281	66,504	272,728	3,440,604
	1996	246,139	215,878	614,335	327,374	1,403,726	20,613,412
Nonconsumptive	1985	45,018	69,449	148,707	300,039	563,213	14,267,213
	1991	68,017	102,205	362,111	511,218	1,018,878	18,103,887
	1996	146,105	218,864	692,734	1,660,136	2,717,839	29,227,888
Total	1985	340,798	456,166	805,181	1,102,577	2,704,722	55,659,765
	1991	388,286	291,455	1,069,164	1,904,118	3,653,023	59,027,316
	1996	826,819	718,576	2,219,290	2,270,161	6,034,846	101,162,130

Source: USDOI 1989, 1993, 1998.

to 1986, but most of the increase in apparent use was on areas that were added to the National Wilderness Areas Protection System since 1971 (USFS 1988b; Darr 1989). Use of these areas before 1971 apparently was not counted as wilderness recreation because the land had not been formally designated as wilderness by Congress. Thus, the apparent increase in wilderness recreation on formally designated wilderness areas over this period may not be real. In addition, data for the national parks indicate an unexplained decline in overnight stays in major wilderness parks since the mid-1970s. This decline, if real, may be response to any number of factors. It may indicate a relative decline in interest in wilderness recreation. Or it may indicate that users are recognizing that overuse of wilderness areas degrades the quality of the wilderness experience. Or it may indicate that limits being placed on wilderness use by the administering agencies are having an effect on use.

The current mix of recreation uses of forests clearly reflects, in addition to a variety of demand factors, the supply of recreation opportunities and charges for their use. Restrictions on use, real and perceived, affect the balance of use between public and private forests, as do the conditions of the forest. Inasmuch as most of the private forests in the region have been logged at least once and are managed fairly intensively for wood products, recreation that requires extensive areas of relatively wild land occurs mainly on public lands. But other kinds of recreation, especially those that involve recreational and off-road vehicles, might be spread more evenly between public and private forests. Changes in forest management brought about by the Northwest Forest Plan will affect the future mix of available recreation opportunities in the region.

Fisheries

Streams that emerge from or run through Pacific Northwest forests support important regional fisheries. Commercial fishing is limited mainly to anadromous species; sport fishing encompasses anadromous and nonanadromous inland fishing. Most of these fisheries depend on cold, clear water. Spawning usually requires silt-free, gravelly streambeds.

With the exception of those for some of the salmon stocks in the Pacific Northwest, data on trends in fish populations are almost nonexistent (Flather and Hoekstra 1989; FEMAT 1993). What data there are do not indicate the degree of dependence of fish numbers on forests.

Commercial and sport fishing have been important economic activities in the region, and anadromous salmonids have accounted for a significant part of the overall fishery. For example, salmon in 1978, a typical year for the period from 1970 to 1986, accounted for about 12% of the weight of commercial fish landings in Oregon and for about 25% of the value (Carter 1988). In recent years, the share of salmon in weight and value has fallen in Washington, Oregon, and northern California fisheries. Salmon's proportion of the total weight of seafood landings fell from 6.6% in 1989 to 4.8% in 1991, while the share of value of total landings fell from 16.8% to 10.8% (FEMAT 1993). Salmon and crab have consistently accounted for more than their share of value relative to weight of total landings.

The value of salmon landings from commercial troll ocean fisheries in the region has varied widely over time. From peaks in the late 1970s and again in 1988, the value of landings in 1992 and 1993 was lower than at any time in the previous 15 years. Recreational catch was also low in 1992 and 1993. The economic impacts of ocean salmon fisheries on coastal communities have been substantial. In 1987, fishing contributed about 11% of the total personal income in an Oregon coastal area made up of five complete counties and coastal portions of two others (Radtke and Davis 1988). The timber industry accounted for about 15% and tourism for about 7% of the area's total personal income at that time.

In 1974, a poor year for salmon, sport fishing accounted for 65% of the total value of salmon from the Columbia River, including commercial, sport ocean fishing, and river fishing. The value of ocean sport and ocean commercial fishing were about equal, but river sport fishing contributed nearly 6 times the value of river commercial fishing (Powel and Loth 1981). In terms of its overall contribution to the economic impact of forest-related recreation, fishing accounted for about 6% of the annual expenditures on recreation on BLM and national forest lands in the northern spotted owl region in 1990 (FEMAT 1993).

State-to-state differences in the role of anadromous fisheries are significant. About 33% of the sport-fishing activity 1975-1977 in Oregon and Washington, but only 4% of the sport fishing in Idaho, was for anadromous fish (Powel and Loth 1981). But cold, clear water is important to the sport fisheries in all three states. Warm-water fishing accounted for only 12% of the sport-fishing activity in Idaho in 1975, while fishing for resident trout in streams and in lakes accounted for 46% and 29%, respectively.

Recognizing the potential effects of logging on fish habitat, the Pacific Northwest states have regulated logging practices in streamside zones in recent years. The intent of regulations in Idaho, Oregon, and Washington is to maintain streambank integrity and cool water temperatures. At least until recently, regulations reduced but did not prohibit tree removals in streamside zones

Mushrooms

Mushrooms are one example of a "special product" of Pacific Northwest forests. Others include decoratives such as floral greens and landscape materials, medicinals and herbs, and foods such as berries (Molina et al. 1997). Together they account for a modest share of the marketed products of Pacific Northwest forests. The use of wild mushrooms is the example chosen here for discussion to represent a broad and varied set of nonwood products of Pacific Northwest forests.

Commercial harvesting of mushrooms provides income for some people in the Pacific Northwest. Many of the favored mushrooms are the reproductive structures of mycorrhizal fungi that have symbiotic associations with tree species on Westside and interior forests. Most of the common mushrooms collected in the Pacific Northwest are mycorrhizal (Molina et al. 1993). Other mushrooms collected are either saprophytes or root rotters — e.g., edible morel (forests and nonforested areas) and cauliflower mushroom (mature conifer forests), which do not form symbiotic associations with tree species. A discussion of the ecological role of mycorrhizal associations is in Chapter 3.

High interest in individual collection of mushrooms in the Pacific Northwest is shown by a large number of amateur mushroom societies. Individual collectors stimulated the regulation of commercial mushroom collection in Washington in response to the increase in commercial mushroom harvesting and to the competition for this unmanaged resource that began in the 1980s. Mushroom pickers in Washington now must buy a license and can be monitored, which provides something of an information base for the activity.

Information on harvest levels and sales for commercial mushroom harvesting is available starting in 1989 and 1990. That information, however, is limited because commercial pickers probably represent only 10-20% of what was actually collected. The data indicate that $652,247 and $1,278,910 was paid to licensed buyers and processors (dealers) for harvested mushrooms in Washington in 1989 and 1990 (Molina et al. 1993). Morel production in Oregon in 1987 was estimated to be worth more than $2.6 million (equivalent to the value of the state's blueberry crop). Matsutake harvesting earned $9-10 million for buyers and dealers in British Columbia in 1988 from sales of mushrooms to Japan.

The sustainability of mushroom harvesting still needs to be determined. That is a real concern, because mushroom harvests in Europe have declined. Some of that decline can be attributed to pollution (Arnolds 1991), but some can also be attributed to changes in land use and in tree-species composition in the forests. Switzerland, Italy, and Germany have regulations that control or limit mushroom collecting in some regions (Molina et al. 1993), especially in some high-elevation forests where symbionts are critical for tree growth. In the United States, only Washington regulates mushroom harvesting, and that is limited to commercial harvesting. Molina et al. (1997) lists and categorizes information and research needs for adequate management of special forest products, including mushrooms.

Water

Water is an important nonwood product of Pacific Northwest forests, but one that received little attention in the Northwest Forest Plan. The Northwest is generally well watered, and water usage is not threatened by limited supplies. Changes in management as a result of adopting the plan presumably will have beneficial effects on the overall average quality of water flowing from the region's forests and some effect on the timing of flows. These effects will be more important locally than regionally.

Supplying residential and community water continues to be an important concern of forest management. Portland and Seattle, as well as many smaller communities in the region, depend on protected watersheds for their water. Protection of these watersheds to ensure high-quality water will continue to be an important consideration.

Overall freshwater use in the Pacific Northwest is projected to remain nearly level from 1995 to 2040. Irrigation use, by far the largest use in the region, is projected to drop, but other uses are expected to increase. Domestic and public use in the region, a relatively small part of the total but one that requires the highest quality of water, is projected to increase by 44% over the next four decades (Brown 1999).

Effects of Changes in Management of Pacific Northwest Forests on Nonwood Products

The Northwest Forest Plan was aimed at maintaining habitat for various species dependent on old-growth forests. But the reductions in Pacific Northwest federal timber harvests as a result of the Plan will also:

- Favor some kinds of wildlife, game and nongame species, over others
- Affect hunting conditions and hunters' expectations
- Improve habitat for anadromous and inland sport fisheries
- Maintain some kinds of backcountry recreation opportunities

Information on the effects of adopting the Northwest Forest Plan on nonwood forest products in the Pacific Northwest is spotty. For example, most of the information on effects on wildlife populations, aside from that concerning the northern spotted owl and other species at risk of extinction, has been with respect to big-game species. One study of the effects of forest structure on breeding birds in the Oregon Coast Range found that habitat fragmentation due to logging had mixed effects on bird populations (McGarigal and McComb 1995). But relating the results of even this study to the changes brought about by the Northwest Forest Plan is somewhat speculative.

Changes in future backcountry forest recreation opportunities on federal land will depend on the rules adopted for old-growth and late-successional reserves, other than designated wilderness areas, for which rules are clear. Rules similar to those that now apply to designated wilderness areas will lead to a set of results that are different from those that will result from rules that are less restrictive. Presumably there will be more opportunities for backcountry and wilderness-type recreation as a result of the cutbacks in federal timber harvests than would otherwise be the case.

Effects of the Northwest Forest Plan on mushrooms and other special forest products and on water flows are also uncertain. Molina et al. (1997) note that the lack of information on the complex biology of managing and harvesting special forest products poses difficulties in integrating their management into broad ecosystem management guidelines.

REGIONAL ECONOMIC EFFECTS

Estimating the regional economic effect of shifts in the proportions of wood and nonwood products resulting from reductions in federal timber harvests in the Pacific Northwest is difficult. For example, the extent to which timber harvests are competitive with or complementary to nonwood products is not clear. In addition, economic data are not collected in a way that allows for ready estimates of income or other measures of economic impacts from the nonwood industries. At best, most estimates of economic impacts related to nonwood forest products are patched-together proxies for direct measures. There is even dispute over the effects on output and employment in the wood industries, for which there are fairly reliable measures (Tuchmann et al. 1996; Haynes and Weigand 1997).

Estimates of such economic measures as "expenditures for wildlife-associated recreation," "value ... paid to harvesters [of mushrooms]," or "yearly recreation benefits" (FEMAT 1993) are almost meaningless by themselves or in the absence of trend information. They usually cannot be compared with standard economic measures of performance for other sectors and, therefore, are not useful in estimating net effects of changes in policies or programs. The basis for estimating economic welfare effects of changes in nonwood products outputs that could be compared with those for changes in wood products outputs is exceedingly weak (Haynes and Weigand 1997).

Even official statistics for well-defined measures can be misleading if not adequately put in context. For example, the value of shipments of the lumber and wood products industry in Idaho, Montana, Oregon, and Washington in 1987 was $16.5 billion (US Bureau of the Census 1991). If the timber-dependent woodpulp and paper industries were added, the total would be well over $20 billion. But, these numbers by themselves can also be highly misleading. How do they, for example, relate to parallel numbers for nonmanufacturing forest-related industries? How much of the value is for shipments outside of the region (export base) relative to that which stays in the region? Do the numbers reflect particular stages of the business cycle? In the absence of answers to these questions, we have chosen not to present further economic impacts estimates.

SUMMARY

The reduction in federal softwood timber harvests in the Pacific Northwest has resulted in a roughly equal increase in softwood timber harvests in the U.S. South and Canada. This has come about in response to normal market forces. The increase in southern timber harvests is being met in part by increases in the intensity of forest management, especially on forest industry land. This increased management intensity potentially will affect some environmental values, such as maintenance of wetland ecosystems and protection of species such as the red-cockaded woodpecker.

Nonwood forest products in the Pacific Northwest for the most part are not competitive with similar forest products from other regions. The extent to which their availability to markets within the Pacific Northwest has been affected by adoption of the Northwest Forest Plan is generally unclear due to lack of information based on research results.

7
FOREST MANAGEMENT AND RURAL COMMUNITIES IN THE PACIFIC NORTHWEST

Rural economies in the Pacific Northwest historically have derived a substantial share of their economic base from natural-resource-based industries, including timber, agriculture, and mining. As with many similar rural economies across the country, they have had trouble adjusting to new circumstances when economic or policy changes affect the availability of the resources that have been important to them. Some 5.6 million Pacific Northwest residents live in the 129 counties classified as nonmetropolitan by the 1990 U.S. Census; that is 46% of the total population and 85% of the 151 counties in the region. Analyses in this chapter are based on county-level data unless otherwise noted. Forests and forest-based industries have been important to many of these nonmetropolitan counties in the region.

RURAL ECONOMIC WELL BEING AND NATURAL-RESOURCE INDUSTRIES

Rural poverty is not unique to the Pacific Northwest. Throughout the United States, nonmetropolitan areas almost always have higher poverty levels than do metropolitan areas (RSS 1993). And although nonmetropolitan areas in the Pacific Northwest have lower incomes and higher poverty rates relative to metropolitan counties, they have fared well compared with persistent poverty areas of the South, Southwest,

Alaska, and the Southern Highlands. Of the 540 U.S. counties that have had poverty levels of 20% or more since 1960, only one of those counties is in the Pacific Northwest (Beale 1993).

In 1990, the median nonmetropolitan family income in the Pacific Northwest was $23,159 compared with $29,481 for metropolitan counties (U.S. Bureau of the Census 1990). Only one other nonmetropolitan region—northern New England—had a higher median family income ($26,051) than the nonmetropolitan counties of the Pacific Northwest. In contrast, the Upper Great Lakes nonmetropolitan family median income was $20,867, and the median income in the Mississippi Delta was $15,532.

Those income figures are mirrored by poverty data. In 1990, 15.3% of the nonmetropolitan Pacific Northwest population lived in poverty while 12.2% of the metropolitan population did (U.S. Bureau of the Census 1990). In comparison, nonmetropolitan northern New England had a poverty rate of 12%, the Northern Plains had 17.3%, the Southern Coastal Plain had 26.4%, and the Mississippi Delta had 34.3%. The reasons for lower income levels and higher poverty rates in rural areas are complex and the subject of considerable study and debate (RSS 1993). One explanation is rural America's dependence on volatile natural-resource markets (Humphrey et al. 1993). Dependence on a few industries for the majority of local economic activity also limits the capacity of rural economies to absorb change.

Two long-term trends in the manufacturing sector are important in the Pacific Northwest rural economies. The first is increasing efficiency in raw material use. The second is increased labor efficiency and declining demand for labor, particularly among the ranks of blue-collar workers, once the mainstay of rural economies. The increasingly competitive world market places greater pressure on U.S. manufacturers to continue such productivity increases (Galston and Baehler 1995).

The lumber and wood products industry in the Pacific Northwest illustrates the effect of the general trend in natural-resource industries. Young and Newton (1980) conclude that there is a "long term trend toward capital intensiveness in the wood products industry. The substitution of capital for labor in production means the loss of jobs." Hibbard and Elias (1993) report that in the late 1970s, 534 timber mills produced 11 billion board feet of lumber and employed almost 200,000 workers. In 1988, 453 mills (15% fewer) produced 16.5 billion board feet (a 50% increase) with 160,000 workers (18% fewer). Between 1979 and 1988, 26,000 jobs in the lumber and wood products sector were eliminated as output rose (Anderson and Olson 1991).

Although there has been a general trend toward increasing labor productivity and consolidating operations across the region, it has not been evenly distributed geographically. Cordray and Goetz (1994) showed that the mills that closed in Lane County, Oregon, were in the rural areas of the county; those that remained open were in the urban areas. Those closures and similar relocation occurred during the late 1970s and the early 1980s as the timber industry sought to reduce its manufacturing and transportation costs.

The estimates of job loss in Washington and Oregon owing to restrictions on federal harvest range from a low of 12,000 jobs (Anderson and Olson 1991) to a high of 147,000 jobs (Beuter 1990). The differences are the result of differing assumptions regarding the time involved and the change in timber supply. Sample and Le Master (1992) compared those and other estimates using standardized assumptions and estimated that employment decline would range between 11,858 and 32,015 jobs from 1991 to 2000. FEMAT (1993) concluded that relative to 1992 levels, its projections of timber harvests for the various options being evaluated and its assessment of regional employment per unit of harvested timber "imply a range of job displacement from 21,200 to 32,000 jobs." That is fewer than 3,000 jobs a year in an economy that is growing by more than 241,000 jobs a year.

Although relatively few jobs are being lost from restrictions on harvesting, it is one of a long string of events that have reduced the importance of the timber industry in the economy of the Pacific Northwest. FEMAT concluded "timber-based employment is apt to be declining under all options considered. . . . [T]he economy of the region as a whole appears to be poised for continued growth. The job loss issue thus becomes more of a distributional nature, with rural communities declining as more developed areas expand."

The next section looks more closely at the specific social and economic outcomes of timber dependency in the Pacific Northwest.

TIMBER DEPENDENCY AND COMMUNITY WELL BEING

The committee was asked to "evaluate to the extent possible the nature of the economic and social costs and benefits associated with alternative management practices." Although the question is easy to ask, it is hard to answer. Few social-impact studies clearly tie social and economic outcomes with specific forest-management practices, such as old-growth harvest rates, the use of clearcutting as a harvest technique, or the relative intensity of silvicultural practices. Several studies have examined the relationship between the percentage of people employed in timber producing and processing industries and indicators of well-being such as income, percent living in poverty, and housing conditions. Counties with a higher proportion of such jobs relative to other counties are referred to as timber dependent.

Heberlein et al. (1994) reviewed eight studies by Drielsma (1984), Elo and Beale (1984), Kusel and Fortmann (1991), Bliss et al. (1992) Howze et al. (1993), Lee and Cubbage (1993), Force et al. (1993), and Overdevest and Green (1994) and presented a meta-analysis of the relationship between varying levels of timber dependence and measures of community well being. The studies covered the Northeast, the Pacific Northwest, the Southeast, and the entire nation as a whole. Those studies reported relationships between the proportion of timber jobs and 134 measures of socioeconomic well being.

The majority of the relationships between increasing timber dependency as measured by the proportion of timber-related jobs and social and economic well-being indicated that well-being went up as timber dependency went down. In most cases, timber dependency seemed to hurt rather than help communities. This analysis found that timber-dependent counties (and by extension, communities) tend to have higher unemployment, lower income, more poverty, and lower levels of education in comparison with counties with greater economic diversity. They also have older, lower-value housing that tends to be seasonal, with fewer new houses. Timber-dependent counties tend to have lower birth rates, higher death rates, greater age dependency, higher infant mortality, and lower growth rates than other counties. Some evidence points to poorer health care, fewer churches and more arrests.

A second set of research linked management variables with social and economic well being. Rudzitis and colleagues have been studying the effects of wilderness preservation on community well being. Wilderness is a special designation of federal land in which management practice is restricted to protect the area from timber cutting and motorized vehicle access. Wilderness designation has been controversial because of its presumed negative effect on local timber harvesting jobs. Thus, lower levels of economic well being might be expected in communities located near wilderness areas. However, the available data show that is not the case. Rudzitis and Johansen (1991) found that the unemployment rate in wilderness counties (counties that either contain or are adjacent to counties that contain federally managed wilderness areas) is well below the national average. Adjacency to protected lands like wilderness actually serves as an attraction for new residents in an area. Rudzitis (1993) later found that wilderness counties showed population growth of 24%, 6 times faster than the nonmetropolitan counties nationally and twice as fast as nonmetropolitan counties in the west (Rudzitis 1993). Wilderness attracts people of greater economic means and thereby increases the level of socioeconomic indicators of well being. The migrants to these counties are more likely than long-term residents to be college graduates and have professional occupations and higher incomes. Four out of five in-migrants to wilderness counties rated scenery, outdoor recreation, and environmental quality as the most important reasons to move (Rudzitis and Johansen 1991). These studies suggest that wilderness and amenity protection can have a positive influence on certain measures of community well being, although in-migration brings its own difficulties (Brown 1993).

Whether county-level data represent communities has been questioned, because county data aggregate data from several communities. Employment and income data, however, are maintained at the county level, and economic regions are generally larger than individual villages, towns, and cities. Also, timber dependency or forest-management effects might be masked by using only 1980 and 1990 county-level census data.

Force and co-workers (1993) examined community-level data over time for a single community (Orofino, Idaho) to explore this concern. They used government, media, industry, and personal sources from as far back as 1920 and showed that social unrest, disasters, and nontimber development have a more profound effect on community well being than do changes in forest harvest and mill production. They also found that the number of employees and churches decreased and the number of arrests increased as harvesting on national forests increased. Those findings are consistent with the results of other county-level, cross-sectional analyses.

Force et al. (1994) extended their analysis to four resource-dependent communities — communities dependent on timber, mining, fishing, and tourism. They examined the joint effects of local resource production (volumes of wood, minerals or fish, number of employees in resource-dependent industries, and product values), local historical events, and societal trends to determine the effects on four indicators of community social change (size, structure, cohesion, and anomie[1]). In only 5 of 20 cases (in a matrix of four social change variables by five communities) did resource production correlate with social indicators. When it did, the effect was sometimes an inverse relationship between resource production and well being. In Prineville, (the timber-dependent community) resource production had no effect on measures of community size, structure, or cohesion beyond the effect of local historical events and societal trends. However, increasing numbers of divorces (one indicator of anomie), did exhibit a direct relationship with increased timber cutting independent of social trends and local historical events. Thus, the in-depth statistical analysis of communities over time also fits the general conclusion that timber dependency is associated with lower levels of certain measures of social and economic well-being.

DIVERSIFICATION IN RURAL COMMUNITIES OF THE PACIFIC NORTHWEST

As industries such as agriculture, mining, and forestry decline as employers and sources of income, economic diversification in rural America has emerged as an urgent need. Diverse economic conditions create diverse opportunities and thus temper the effects of timber industry fluctuations on local communities (Kusel and Fortmann 1991).

Over the past 20 years in the Pacific Northwest, rural communities have become much less dependent on timber, mining, fishing, or agriculture (Anderson and Olson, 1991). They increasingly benefit from a mix of extractive industries, light manufacturing, retirement, residential, service, and recreation sectors. In Oregon, for example, the lumber

---

[1] The state of alienation resulting from the loss of social stability.

and wood products industry now represents only 5% of total employment. Many of the communities in the Pacific Northwest started off as timber towns, but have made a transition to an entirely different economy (e.g., Leavenworth, Washington, and Sisters, Oregon). Many more have seen their local economies diversify.

The signs of economic diversification evident in economic trend data can be easily overlooked, because the logging culture permeates the Pacific Northwest. Log trucks and mills are obvious, but small businesses and home-based employment are not. Cultural symbols are infused with the logging culture: the community college in Aberdeen, Washington, is called the "Home of the Chokers" (a choker is a common piece of logging equipment), and the Libby, Montana, high-school teams are the "Loggers." But self-employed service workers and transfer payments of retirees are not visible, although their incomes contribute significantly to diversifying rural communities. It is those dollars that are making a significant contribution to rural communities as they diversify and grow in new directions (Power 1992).

Rural economic diversification in the Pacific Northwest is most directly affected by the availability of transportation and proximity to population centers. Rural northwestern communities lying along Interstate 5 — running from the Canadian border to Los Angeles — have enjoyed substantial economic growth in recent years, in businesses ranging from computer hardware and software to telephone-switching centers and trucking companies. The National Association of Counties (1993) notes that "Mill City is becoming a 'bedroom community' since many who lost local jobs are traveling 15-50 miles to jobs." Regression models showed a positive effect of commuting on household income net of other factors (Heberlein et al. 1994).

Kusel and Fortmann (1991) note that "locally-based ecosystem restoration is a growing phenomenon in California," with state support expanding to provide local jobs. Local tree-planting and restoration organizations have grown up based on funding provided by such programs; in some areas, they have become an important source of employment. In addition to forest restoration, the development of nontimber forest products has shown increasing influence in the regional economy of the Pacific Northwest. In their 1991 study of special forest products in the coastal part of the region, Schlosser et al. (1991) found that "floral greens" has expanded to become an important regional industry. Its 1989 value-added contribution to the gross national product was estimated to be $80.5 million. Mushroom harvesting and growing and berry harvesting are other examples of diversification.

Many forest communities provide attractive sites for tourism and recreation. As the importance of extractive industry declines, Pacific Northwest communities are looking toward tourism as a way to bolster their economies. A strong tourism industry could strengthen the local economies by providing jobs with low, entry-level skill requirements, potential for upward mobility, and local ownership and control (McCool et al. 1995). However, tourism by itself is not a substitute for timber industry jobs. Annual incomes in tourism sector jobs often are inferior to traditional rural employment, largely because tourism-related jobs are seasonal (Power 1996). In his comprehensive study of Montana's tourism industry Barrett (1987) concluded that "in almost every aspect measurable by census data, tourism employment is substantially inferior—not only to employment throughout Montana's economy but to comparable retail trade and personal services" (Barrett 1987). Such aspects include highly erratic seasonal employment, a high percentage of women and young adults in the work force, and an average income half of the Montana average. Those characteristics can also be strengths, however, for people not seeking full-time jobs or those who are just entering the work force.

> *As the importance of extractive industry declines, Pacific Northwest communities are looking toward tourism as a way to bolster their economies....However, tourism by itself is not a substitute for timber industry jobs.*

Many rural communities have tried to entice new manufacturing companies to their areas to replace lost manufacturing jobs, but many communities are too small and too far from a metropolitan area to attract large-scale industry. A substantial number of firms in rural areas have payrolls of fewer than 100 employees (Castle 1993), but those firms account for the fastest-growing economic sector in nonmetropolitan areas and in the nation's economy as a whole. Kusel and Fortmann (1991) note that because small businesses are location specific, they are most likely to contribute to community well being by using local expertise and management and are more likely to invest in sustainable management of local resources. New small businesses add to the capacity of rural communities, but mill and forest jobs often add only income.

Mobility and the information age are allowing more people to move to rural communities. Cromartie (1994) notes the growing importance of the private service sector in the West: "Much evidence indicates that nonmetropolitan areas in the West are beginning to benefit from the location of high paying, producer services." High-end producer services, such as engineering, accounting, and legal services, which were once found exclusively in metropolitan areas, are increasingly moving to nonmetropolitan areas. Egan (1994) quotes a mill owner in Medford Oregon "people moving to southern Oregon from California are not all retirees, as the stereotype has it. They are bringing in jobs with them."

IN-MIGRATION

Although many nonmetropolitan counties lost population in the 1980s, the elderly are an important group of in-migrants to rural communities. The number of elderly in the United States has grown from 16.2 million in 1960 to 31.1 million in 1990. For the past 3 decades, older people have been leaving metropolitan areas for nonmetropolitan areas, and as the nation ages, rural areas are most affected. Fuguitt and Beale (1993) show that the North Pacific coast region had the third highest rate of in-migration of elderly among any of 26 U.S. regions (the Southwest and the Florida peninsula were first and second). Elderly who migrate from predominantly urban areas to the Pacific Northwest have expectations for a high quality of life, and access to amenity-based resources are a pre-eminent value sought (Fuguitt and Beale 1993, Salazar et al. 1986).

In-migration by retirees brings new income into rural economies in the form of retirement income earned elsewhere. The growing importance of transfer payments in nonmetropolitan areas is changing the traditional dynamic of employment stability in the region and contributing to rural economic diversification. Over the 20 years from 1971 to 1991, nonlabor income increased from 26% of total personal income to 34%. In-migration of retirees also leads to new jobs in the residential, health-care, retail, and amenity-related sectors. They can infuse new ideas and vitality into a community through volunteer work and entrepreneurial activities. This population creates the demand for high-quality community infrastructure, such as hospitals, transportation, and recreation activities.

In-migration also has negative effects. Property taxes are likely to go up as sales increase and new houses are built with outside money. That might reduce the housing available to some residents. The new jobs that are created tend to be service oriented and low paying. There may also be overload on roads, hospitals, fire departments, and sewage treatment plants. Most dramatically, urban and nonurban values might clash, especially where new residents are not sympathetic to the need for local extractive industries (Price and Clay 1980; Blahna 1990). The focus on private property rights might take away common areas that are taken for granted by rural residents (Brown 1993).

POVERTY AND PLENTY—ANGER AND HOPE IN THE PACIFIC NORTHWEST

The economy is growing, and rural communities are diversifying and revitalizing as they rely less on an extractive economy. Nonetheless, strong complaints come from the rural areas of the Pacific Northwest. Mills have been closed and jobs lost to increases in efficiency and productivity. Those are changes caused by a particular set of political and economic decisions. The losses occurred long before owls or environmentalists were a dominant force in the political process. Timber management as it has been practiced in the past century has not had an appreciable effect on the well being of rural people. National and world economic forces have had a more direct bearing on employment than has timber supply (Waggener 1990).

Some jobs have been lost due to recent change in forest policy, lawsuits, and restricted production. Although not large in absolute or even relative terms, those job losses create hardship for the individuals who lost them. Persistent poverty, historic job loss, and current job loss are all having real effects.

Cultural and lifestyle images are probably more important factors influencing people's perceptions of the significance of loss of timber-based employment. Rural culture has emphasized and glorified some extractive occupations—the farmer and the logger, among others. Forestry is a high-visibility occupation and culturally important activity. The pioneer tradition is only several generations removed from present day in the region. Many feel that traditional timber management built the Northwest, and this legacy is being threatened. Retraining for jobs at the same wages, trading a logger's suspenders for a desk job is a loss of identity and status in a rural community. Another response to job loss has been movement out of the community by families who have lived there for generations.

> *The effects on overall regional employment and income of the changes in Pacific Northwest forest practices over the past decade have been relatively small.*

SUMMARY

The effects on overall regional employment and income of the changes in Pacific Northwest forest practices over the past decade have been relatively small. Although some communities that were dependent on federal timber sales have suffered, the combination of a vibrant regional economy and federal assistance to affected people and communities have ameliorated possible impacts. In addition, a rise in the relative importance of income derived from the recreation sector and transfer payments has benefitted rural communities.

The importance of timber-related income in the region's economy has continued to fall as the economic base has diversified and forest products firms have become more efficient. Nevertheless, timber will continue to play a role as a source of income to the region, one that deserves to be weighed in decisions about forest management practices.

8
A FRAMEWORK FOR SUSTAINABLE FOREST MANAGEMENT

Introduction

Each of the goals identified by the committee (sustain viable populations of indigenous species, maintain properly functioning ecological processes, meet human needs for forest commodities, and satisfy cultural and aesthetic values) depends on the functioning of key ecological or ecosystem processes. Maintaining these processes is being adopted as the fundamental goal for sustainable forest management. Not all goals can be maximized concurrently; therefore, balancing partly incompatible goals through forest-management practices is the major challenge facing forest managers. Although much of the debate over forest management has focused on old-growth stands in forested landscapes and on effects of clearcut harvesting on biodiversity, sediment flow, and fisheries, forest-management practices have been in constant transition. Partial-harvest practices have become more widely used, newer harvest and replanting practices have lowered effects on soils and streams, and landscape analyses are being developed to address issues of representation and connectedness of key habitat types. As they have in the past, forest-management practices will continue to change in response to new information and changing societal values.

ELEMENTS OF FOREST MANAGEMENT

Forest management has come to represent far more than simply logging and silviculture aimed at fiber extraction. It is the variety of actions applied to forested landscapes to achieve the goals described above. In more specific terms, the committee viewed forest management to comprise four elements:

Allocation: the allocation of land to particular uses or combinations of uses.

Rationing: rationing or scheduling use, such as levels of timber harvest, permitted recreation use, and bag limits on game and fish.

Harvest: determining how to harvest or take forest products, which are often related to plans for the next cycle of management.

Investment: investment in the productive resources of the land, including protection against wildfire, insects, disease, and other threats; holding trees as timber and resource capital; management practices such as fertilization or irrigation intended to increase value of forest products; and roads and infrastructure.

Land Allocation

At the regional level, basic patterns of forest land ownership in relation to other uses such as development and agriculture have already been largely made. Congress has used, and continues to use, its plenary authority over the federal lands to designate national parks, wilderness areas, wild and scenic rivers, national recreation areas, and others. These decisions, while having a broad regional impact, have usually been made individually with little relation to each other or to other land categories, and with little attention to a regional scheme.

The ecological consequences of regional land allocations among ownerships and jurisdictions have become increasingly clear. At the extreme, the so-called "checkerboard lands" have allocated uses and defined boundaries with no regard for ecological processes such as hydrology, natural disturbance, or wildlife behavior. Allocation of uses within particular classes of land ownership (e.g., federal, state, private, etc.) is generally made with little cognizance of ecological processes occurring at larger scales or of surrounding land uses and objectives.

Although Congress has been specific in setting uses of some designated federal areas, it has left allocating uses on the bulk of the national forest and BLM land up to the land-management agencies. The Forest Service and BLM use a planning process for allocating land uses under broad multiple-use guidelines that give the agencies wide discretion in allocating federal lands to specific uses or mixture of uses. This planning process identifies streamside and scenic influence zones, special wildlife habitat, and intensive recreation use areas as well as timber management and harvest areas. Consideration is given—is required to some extent—to landscape- or watershed-level considerations but for the most part, those plans are for federal lands only. Although they may be based partly on recognition of prior decisions on intermingled private lands—e.g., patterns of previous logging in a watershed—decisions in federal land-use plans only control actions on federal land.

At the private and state levels, the Endangered Species Act has affected land allocation by means of habitat conservation plans, regulatory actions, and state programs and agreements. A notable example is the Oregon Plan for Salmon and Watersheds, which has as a goal to restore populations and fisheries to productive and sustainable levels that will provide substantial environmental, cultural, and economic benefits. Only on the larger private ownerships, usually those of forest industry firms, can private forest owners make decisions that encompass entire landscapes or watersheds. But, even the largest private forest ownerships in the region are usually fragmented and intermingled with those of other owners. Each owner can allocate its forest lands to uses and intensities of use that depend on the owner's objectives. Land-use controls, such as those of the Oregon Land Conservation and Development Commission, may limit an owner's ability to change from forest to nonforest uses, but they typically do not require the owner to devote the land to particular forest uses.

The major land allocations affecting forest land in the region are those in the Northwest Forest Plan. Those allocations of federal forests override previous allocations made in the planning process, but not those made by Congress. The allocations create a set of reserved old-growth forest areas, riparian reserves, and adaptive management areas on federal forests in the spotted owl-region, each with a specific set of management and use restrictions and guidelines. These old-growth reserves cover some 7 million acres, about 30% of the total area of federal forests in the spotted-owl region and roughly the same total area of forest that was already in congressionally withdrawn areas. In addition, the Northwest Forest Plan put about 9% of the federal forests in riparian reserves (pending watershed analyses) and 6% in adaptive management areas (FEMAT 1993). Whether parallel allocations will be extended to Eastside federal forests as a result of current studies is uncertain at this time.

Rationing Uses

A basic forest-management and conservation method is to ration or schedule use over long periods to avoid depleting the ability of the forest to maintain productivity. This is common practice not only among government agencies, but also for forest industry firms in managing for timber production. The main management laws guiding the Forest Service and BLM require sustained yield (high-level annual or periodic flows) of all forest resources over time from federal forests. The Forest Service has interpreted this as meaning "nondeclining-even-flow" of timber from national forests.

The idea of harvest scheduling is also captured in the concept of "bag limits" for fish and game, as well as in the "carrying capacity" of rangelands. These are thought of as maximum sustainable levels of harvest in view of the availability of food and habitat for the species in question. For fish and wildlife, bag limits (population management) for all lands are set and enforced by the states. The level and timing of uses should be determined by an understanding of the capacity of forested ecosystems to produce desired goods and services sustainably as well as by market demand. Such "supply-side" rationing has been the stated goal on public and many private lands.

Forest-management agencies and most forest-industry firms take a relatively long-term view and plan for relatively even and sustained annual timber harvests. Private forest owners other than forest-industry firms are more likely to vary timber harvests based on market conditions and other factors, and they are unlikely to be committed to specific long-term harvest schedules. Timber-harvest schedules for a national forest or industrial forest are based on the size of the forest, the existing timber inventory, productivity of the land, expected investments in growing timber, and expected revenues, all over a period of as much as several decades. In the case of industrial forests, market strategies and the expected availability of timber from other owners may also play a role in setting harvest schedules. Rationing or harvest scheduling is usually done for individual ownerships, or in the case of federal lands, for fairly large units such as national forests. Rarely is such rationing done across ownerships or jurisdictions. To the extent that an individual ownership or national forest encompasses a landscape or watershed or larger area, the rationing can be said to be done on a landscape/watershed or even a regional basis.

The rotation age or time between successive cuts is an important element in setting timber harvest schedules. This could be the age at which the maximum total fiber is available for harvest, but today it is often the age at which forest stands are judged to be "financially mature." This is usually taken to mean the age at which the cost of holding trees for another period of time exceeds the increase in value that is expected over that time period. Financial maturity is influenced by the rate of increase in the volume of fiber or merchantable wood, the value of the trees, potential earnings from alternative investments (opportunity costs), and expected changes in real prices for wood. The importance of these factors varies depending on owner's objectives, expected markets, site productivity, and alternative opportunities for investing the money currently represented in the standing trees. No single rotation age fits all goals or circumstances, but the financial rotation age is generally considerably lower than the age at which the maximum total fiber is present. For many private owners in this region, planned rotation ages are 40-60 years, while the goal for public forests is generally longer.

Over the past 60 years, the age structure of Pacific Northwest forests has been changed by logging, but generally with an eye ultimately to balancing the annual timber cut with annual growth over the long term. In the process, most of the old-growth was cut, the overall average age of the forest was reduced, and timber growth rates have increased. Although timber harvest schedules were not set jointly for public and private forests (except in the special case of the Cooperative Sustained Yield Unit in Shelton, Washington), expected harvests from federal and other public forests surely had some influence on harvest schedules for private forests.

Scheduling decisions on one category of land ownership clearly influence demand and, therefore, decisions regarding the rationing of uses on other lands. The harvest schedules under the Northwest Forest Plan recognized the age structure of timber available for harvest. The plan left available a significant volume of timber outside of the late-successional and riparian reserves for planned harvesting. It was specifically planned to provide for most of the annual timber harvest until younger stands were old enough to be cut. In addition, a considerable volume of young timber was placed in the late-successional reserves that was to be available for certain kinds of commercial thinning. And habitat conservation plans (HCPs) implemented under the aegis of the Endangered Species Act that limit or eliminate harvest activities from particular public or private lands will likely influence scheduling decisions on lands not included in the HCPs.

Harvesting

Logging is the most visible and probably the single forest-management action having the greatest impact on forests and their ecological processes. From the 1940s through the 1980s, forest cutting on managed federal lands in the Douglas-fir region was accomplished with dispersed cutting units of approximately 40 acres in size. The resulting landscape was composed of many small forest patches and young plantations with extensive areas in altered microclimates, maximum edge habitat, and minimum interior habitat (Swanson and Franklin 1992). This dispersing pattern was a stark contrast to the practices on private lands where progressive clear-cutting of large areas was employed.

Among harvest methods, clearcutting in which all overstory trees are removed in a single operation has attracted the greatest attention. Reforestation of clear cuts is obtained by artificial planting, natural seeding from adjacent stands or from trees cut during logging. Seed tree harvests involve removal of most of the overstory trees in a single operation, leaving behind singly or in groups a small number of seed producing

> *Logging is the most visible and probably the single forest-management action having the greatest impact on forests and their ecological processes.*

trees. In shelterwood cuts, the mature stand is removed in a series of cuttings that extend over a relatively short period of the rotation. An even-aged cohort of juvenile trees is thus encouraged in the shelter of uncut trees over the harvest period. All of these harvest methods are associated with even-aged forest management (Smith 1986).

Uneven-aged forest management is used to maintain stands of trees with a full range of age classes up to the rotation age. It is generally used when the most commercially important species are relatively shade-tolerant and can grow under the shade of older trees. Trees are harvested as single, scattered individuals or in small groups at relatively frequent intervals throughout the stand and continuous regeneration of new trees is encouraged.

Although the decision to harvest particular stands is generally a part of harvest scheduling for entire public and forest-industry ownerships, the choice of logging methods and associated regeneration actions (e.g., site preparation, planting, and vegetation control) is typically made on the basis of conditions specific to each site or stand. Even-aged management is suited to regeneration of relatively shade-intolerant tree species such as pines and Douglas fir, whereas single tree harvest and uneven-aged management work best for species such as hemlock and true firs which can establish and grow in partial shade. Even-aged systems are most widely associated with single-species stands; uneven-aged systems are more frequently used for multispecies management (Smith 1962).

Even-aged management initiated with clearcutting the forest overstory is generally used on industry and, at least until recently, on most federal lands on the Westside. Nonindustrial private forest owners are more likely to use some form of uneven-aged management or multi-entry even-aged management where overstories are partially removed (partial cutting) on the Westside, although clearcutting is still generally preferred. Partial cutting is occasionally used on all ownerships in other parts of the region.

Typically, logging on steeper slopes (grade more than 30%) is done using cable systems to suspend logs and move them from the stumps to the loading deck. In especially sensitive areas logs may be transported by helicopter or balloon, although these techniques are comparatively expensive. On less-steep terrain, logs are moved with rubber-tired or tracked ground skidding machinery.

The usual practice following harvesting on federal and state lands, forestry industry lands, and some nonindustrial private lands is to do some form of site preparation prior to planting or natural regeneration. Depending on the character of the forest, type of harvest, tree species wanted in the subsequent forest, and condition of the site following logging, site preparation can include broadcast burning of logging slash and undergrowth, piling logging slash with or without subsequent burning, scarification of the soil or killing brush with herbicides. Inasmuch as state forest regulations in all of the Pacific Northwest states except Montana require that logged areas be substantially regenerated to commercially desirable trees, site preparation is an economically integral part of timber harvesting.

Silvicultural practices in the Pacific Northwest have been the subject of a great deal of scrutiny, and many new techniques and innovations are becoming evident. Among these are variable retention harvesting (Kohm and Franklin 1997), biodiversity pathways modeling, and increased rotation times. As use of new techniques becomes more frequent, they are being evaluated for their ability to achieve environmental objectives in managed stands and landscapes. Silvicultural approaches have been evaluated from the standpoint of providing wildlife habitat, plant species diversity, and aesthetic values (McComb et al. 1994; Hansen et al. 1995; Boyle et al 1997).

Franklin et al. (1997) reviewed alternative approaches to timber harvest that may be more accommodating to concerns about the conservation of biodiversity or key ecosystem processes. Although longer rotations certainly provide benefits in this regard, those benefits can be enhanced considerably if longer rotations are coupled to a "variable retention harvest system." This system is based on the strategy of retaining structural elements of the harvest stand—woody debris, snags, seedtrees and shelterwood for at least the next rotation. Franklin et al. (1997) argue that retention of structural complexity serves three major purposes, "lifeboating," "structural enrichment," and "enhancing connectivity." Lifeboating refers to retained structures that provide refugia for biological elements that might otherwise be lost from the area. By structural enrichment they refer to the fact that the retention of structural complexity accelerates the eventual reinvasion of harvested stands by species lost at harvest (e.g., Carey 1995). Retention

A Framework for Sustainable Forest Management

of complexity diminishes the extent of fragmentation of habitat created by harvesting by enhancing among stand connectivity on managed landscapes.

Implementation of specific prescriptions for harvest retention is clearly limited by existing knowledge of the importance of particular structural elements in the diversity of Pacific Northwest forest types. That said, Franklin et al. (1997) argue much information does exist upon which to inaugurate such programs and that they should in any case be coupled to carefully designed monitoring programs and an adaptive management philosophy.

Investment

Forest-management investments include investments in protection (from wildfire, insects, diseases, and other depredations), in various management practices (e.g., planting trees), in infrastructure such as access roads and recreation facilities, and in developing management information. The major investment is usually in the trees themselves, whether they are held for timber production or for other purposes, and it increases with age and size of the trees. As a result, most forest management is very capital intensive. Federal forests typically carry much heavier timber inventories — and thus have greater investments in timber — than private forests in the region. Timber inventories (on a per acre basis) on federal lands are often two to three times those for privately managed lands. This has been identified by some as an example of "inefficient" management of federal timber (Clawson 1976), but different management objectives may account for the differences.

Forest-management investments are often thought of as applying only to growing and using timber, but investments are also required to provide nonwood products such as wildlife. Such investments include the opportunity cost of holding timber inventories to protect riparian and special wildlife habitat, to provide for landscape and other amenities that are attractive to both active and passive users of the forest, and to provide for the production of such miscellaneous products as Pacific yew and mushrooms. And just as there are opportunity costs in holding valuable timber off the market in order to get these nontimber benefits, there are opportunity costs associated with losing nontimber benefits in order to harvest timber and get its economic value. Those costs apply to both public and private forest owners.

Examples of forest investment

Postharvest Site Preparation. The objectives of site preparation include eliminating logging slash and other debris, reducing competition, modifying animal habitat, preparing a mineral seedbed, mitigating compaction, improving drainage, creating a more favorable microsite for seedling establishment, and controlling disease. Depending on the character of the forest, type of harvest, tree species wanted in the subsequent forest, and condition of the site following logging, site preparation can include broadcast burning of logging slash and undergrowth, piling logging slash with or without subsequent burning, scarification of the soil, killing brush with herbicides, and application of fertilizers.

Reproduction. The next stand of trees may be naturally seeded from nearby trees or planted with seedlings, in which case reproduction costs are quite low. Naturally seeded sites typically have very high stem densities and often require thinning at various stages in stand development. As stands develop in stature more of the thinned material can be sent to market, compensating for thinning costs. In many areas, particularly on large cuts or where natural seed rain is limiting, stands are regenerated by artificial planting. In many areas genetically improved seedling stock derived from parent trees selected to favor rapid growth or desired wood qualities is favored (Farnum et al. 1983). Planting of trees at lower densities significantly accelerates the establishment of the stand in the first five years after cutting.

Intermediate Silvicultural Investments. During the growth of a forest stand, a variety of treatments may be applied aimed at improving the existing forest stand, regulating its growth, or providing early financial returns. Such treatments include pre-commercial and commercial thinning, improvement cuttings, pruning, release operations, and fertilization. Plantations and naturally-seeded stands typically have stem densities significantly higher than will survive to maturity or that are desirable for optimal tree growth. A proportion of the trees in the overstory are cut at an early age (typically 8-12 years) to thin tree density and accelerate growth on the remaining trees. Historically, thinned trees have been left on site to decompose, but as stands mature some of the thinned material may be merchantable. Pruning of the lower tree branches at a young age is used in some cases to enhance future log value by providing knot-free wood.

As in agricultural crops, research has shown that augmenting the soil with fertilizer can significantly increase tree growth (Wild and Breeze 1981). Most commonly, nitrogen fertilizer in the form of urea pellets is applied from the air to the site approximately every 5 years. No other type of fertilizer is widely used. Where insect pests present serious risks, pesticides may be applied from the air.

At nearly all stages of stand development protection from wildfire is practiced, particularly in more arid regions. This may take the form of investment in personnel and machinery necessary to suppress fires or maintain protective structures such as fuel breaks and fire lanes. Increasingly, emphasis is being placed on the management of fuels by means of prescribed fire or brush removal.

Gravel and dirt roads are the primary delivery system for logs from forest to mill. The vast forest road network that currently exists in the region represents a major technical and financial commitment of federal and state agencies and private landowners.

Positive and negative incentives to forest investment

Investment decisions are made in response to incentives and disincentives. Incentives might include increases in timber prices or the availability of tax breaks or cost-sharing by the public. Disincentives, such as uncertainty about the future, can reduce investments. For example, early in this century, perceived and real risks associated with wildfire and forest pathogens and pests were strong disincentives for forest investment on private lands (Clawson 1979). Technological advances such as fire-suppression techniques and pesticides provided opportunities to reduce such risks and incentives for forestry investments across a range of ownerships. Ironically, some of these technologies (e.g., fire suppression) have led to landscape changes that have increased risk and uncertainty and complicated investment options in areas such as the Eastside of the Cascades. Positive and negative incentives are influencing forest-management investments in the Pacific Northwest. The reduction in federal timber harvests as a result of the Northwest Forest Plan is leading to increases in timber prices, which in turn lead to more intensive management of private forests for wood production.

Uncertainty about Endangered Species Act requirements is apparently creating a perverse incentive for private owners, leading to premature timber harvesting and reduced management. In particular, uncertainty over whether disruption of spotted owl habitat on private land is prohibited is reportedly leading some private owners to eliminate such habitat by logging before presence of spotted owls can be verified. Others, mainly large landowners, have had "habitat conservation plans" accepted by the U.S. Fish and Wildlife Service. For example, in return for some relaxation of restrictions on harvesting timber near spotted owl nests, the landowners agree to long-term habitat improvements (Weyerhaeuser 1994). That kind of a response involves substantial investments not only in preparing materials for the permitting process, but in the modifications of timber management programs. The latter costs include the opportunity costs of postponing timber harvests and reducing potential timber yields, and the direct costs of using more costly management and logging practices.

Investments on private forests are guided largely by the marketplace—expected costs and returns—as bounded by regulations and supported in some cases by public incentives. To some extent and in some places, these may be guided in part by arrangements among stakeholders, such as the Timber-Fish-Wildlife agreement in the state of Washington. Such investments can be viewed as conforming to regional or landscape level goals. On public forests, especially federal forests, investments are much less responsive to market incentives. Nevertheless, costs, including the opportunity costs of reserving land for specific purposes, are a vital issue.

MEETING THE GOALS OF SUSTAINABLE FOREST MANAGEMENT

During the first decades of this century, principles of "sustained yield" (Pinchot 1907) and "ecological forestry" (Toumey and Korstian 1937) were central as forest managers in North America began to appreciate the need for and value of reforestation and planned scheduling of management activities. As demands for forest resources increased and became more complex, "multiple-use forestry" became the buzz-phrase of choice, particularly on public lands. Today, we speak in terms of "sustainable forest management" or "ecosystem management," recognizing not only the complexity of forest-management goals, but the complexity of the systems necessary to meet them.

With 100 years of experience, forest managers have become adept at managing at least one aspect of a biological system according to a well-defined set of scientific principles and restricted economic and ecological objectives. The forest-management system has reasonably mastered the skills of managing over large spatial scales and learned routinely to make decisions and investments that will be realized only in temporal time scales of fifty or more years. Though management objectives may not always have been targeted correctly, the forest-management system as a whole has adapted to new knowledge and needs as they have become understood (Swanson and Franklin 1992). The evolution to management for sustained yield in the 1950s is now followed by transition to ecosystem management in the 1990's (Salwasser 1990).

Learning has not always been smooth, and changes in forest management have often been fostered by public controversy (Chapman 1931; Swanson and Franklin 1992). Management agencies often lack systematic plans for learning, which should include prioritized listings of identified uncertainties, methods for reducing important and tractable uncertainties, procedures for evaluating existing actions, and mechanisms for retaining new knowledge in the memory of the institution (Hilborn 1992; Christensen et al. 1996). Incorporating large-scale and long-term ecosystem considerations in forest-management decisions generally requires changes in the behavior of the institutions — the market, land ownership, laws and regulations, agency responsibilities — that frame these decisions. Institutional barriers to learning can limit our capacity to change (Senge 1990; Lee 1993). The existing institutional framework is unlikely to be responsive to some of these challenges.

The following section presents a framework for sustainable forest management, with particular reference to the challenges discussed in earlier chapters. The means by which forest management can be made adaptable and accountable in the face of inevitable uncertainty and surprise is discussed. Finally, the challenges and opportunities for resolution of conflict and implementation of management policies across spatial scales ranging from individual ownerships to the entire region are considered.

A FRAMEWORK FOR SUSTAINABLE FOREST MANAGEMENT

Over the past several years, sustainable forest management or ecosystem management has been proffered as the fix for the variety of natural resource challenges facing the Pacific Northwest and elsewhere (e.g., FEMAT 1993; SAF 1993; Slocombe 1993; Gordon 1994; Grumbine 1994; Christensen et al. 1996). However, the need to manage in the context of entire ecosystems rather than focusing only on selected parts of ecosystems has been recognized in the United States for more than half a century (Slocombe 1993, Grumbine 1994). There is still much confusion about what ecosystem management means: to some, it is leaving more snags and logs in clearcuts, to others it is mimicking natural patterns at the landscape scale, and to yet others it is a comprehensive approach that recognizes structures, processes and interrelationships that span a wide range of scales from global to microscopic. All commentaries on this subject have emphasized the importance of reconciling the complex human needs and interests as components of virtually all ecosystems, while recognizing that sustained provision of natural resource commodities and amenities depends on the maintenance of biological diversity and ecosystem processes. Key elements of sustained forest management or ecosystem management (Grumbine 1994; Christensen et al. 1996) are described below.[1]

Operational Goals

Goals should be formulated so as to provide benchmarks for measuring the success of management policies and practices. Operational goals

---

[1] Adaptive management and monitoring are usually included in lists of key elements of ecosystem or sustainable forest management but are here discussed under the heading "Making Management Adaptable."

A Framework for Sustainable Forest Management

pay special attention to ecosystem structures and processes needed to sustain provision of natural resource goods and services.

The goals of forest ecosystem management cannot be stated simply in terms of the commodities or amenities we would like to extract from such systems. Rather, management must acknowledge the importance of basic ecosystem structures and processes that sustain natural-resource goods and services. Given the complexity of ecosystem structure and processes, the setting of operational goals is by no means a simple matter. Nevertheless, the elements of ecosystem functioning discussed in Chapter 4 provide a framework for setting such goals.

Managing in Context and Across Scales of Space and Time

It is in the context of the unique circumstances and objectives of particular places that ecological principles are successfully applied to practical problems. The spatial and temporal context for management decisions should match the scales of ecosystem processes critical to sustainability. The connectivity of the landscape and the reality that actions at one location may influence ecosystem properties and processes elsewhere should be an integral part of management planning. Strategies must be developed to reconcile the incongruity of political jurisdictions and ownership boundaries with ecosystem processes.

Issues of scale head almost everyone's list of forest ecosystem management objectives, and for good reason; successfully achieving any set of management goals requires that each goal be addressed at appropriate spatial and temporal scales. For example, sustaining viable populations of many species requires maintaining sufficient habitat at the regional scale. It also requires a long-term view that foresees inevitable habitat loss and provides for buffering and replacement. Sustaining the potential for sites to produce wood requires

- A landscape perspective because landscape patterns influence the spread of disturbances.
- A local ecosystem perspective because the structure of ecosystems influences their stability.
- A microscopic perspective because nutrient cycling, hence forest

productivity, requires healthy populations of microorganisms and microinvertebrates.
- A temporal perspective because actions that increase tree growth in the short run could in some cases decrease it in the long run.
- A global perspective because some factors that have a direct bearing on tree growth, such as climatic fluctuations and thinning of the ozone layer, are global processes.
- A perspective on complexity, or the realization that the dynamics of nature emerge from interactions between management and numerous natural processes occurring within and across these various scales.

Although forest management has commonly dealt with relatively large scales in making decisions about land allocation, management of multiple use lands, and in particular silvicultural strategies, have focused primarily at the stand level. In effect, the forested landscape was divided into two distinct parts: reserves and areas managed primarily for wood fiber (Perry 1994). Many ecologists and conservation biologists have concluded that sustaining biological diversity in the long-run is not possible solely in a system of reserves (Noss 1983; Walters 1986; Perry and Maghembe 1989; Maser 1990; Norse 1990; Franklin 1991a; Salwasser 1991; Kessler et al. 1992). That recognition has at least two implications for the scale at which forest-management decisions should be made. First, reserves must be located using criteria in addition to their aesthetic worth and low commercial timber value. Second, stand management, and especially silvicultural decisions on multiple use lands must consider goals and context at larger scales. The fact that a growing number of subregional management plans (i.e., smaller scales than FEMAT and PACFISH) cover spatial scales ranging from tens of thousands to hundreds of thousands of acres is encouraging.

APPROACHES TO MANAGING FOR DIVERSITY: GENERAL CONSIDERATIONS

A primary challenge for forestry in the Pacific Northwest–indeed for forestry throughout the world–is translating the guiding principles embodied in ecosystem management into specific practices on the ground and developing tools to assess how successfully goals are being met. Considerable progress is being made along these lines and, although the science and art of managing for sustainability is in its infancy, the Pacific Northwest has been and remains a leader in developing and proving new techniques.

> *A primary challenge for forestry in the Pacific Northwest-indeed for forestry throughout the world-is translating the guiding principles embodied in ecosystem management into specific practices on the ground and developing tools to assess how successfully goals are being met.*

As discussed earlier, one of the primary lessons of modern ecology is the seamlessness of interactions in space and time. One implication of this is that landscapes and regions must be considered as interactive wholes rather than simple collections of individual stands. Nevertheless, when formulating management strategies at the level of landscapes and regions, it remains useful to distinguish land-use types that play different functional roles and that may or may not be tied in perpetuity to a fixed area. In what they called the triad approach, Seymour and Hunter (1992) distinguished three land-use types that, in theory, could coexist at some level within a region without compromising the goal of sustaining biological diversity: (a) reserves, (b) lands managed using "new forestry" techniques (i.e. where maintaining diversity and ecosystem function takes precedence over maximizing timber production), and (c) intensively managed lands (where the emphasis is on producing fiber). From the standpoint of maintaining biological diversity, the triad approach (or any other that recognizes different land-use types) reflects the fact that not every piece of ground must function as suitable habitat all the time to maintain viable populations. The key scientific questions center on the proportion and spatial arrangement of the three types that give a reasonable probability of maintaining diversity through time, the time requirement implying that dynamics-disturbance and recovery-must be taken into account. This is the approach proposed in FEMAT (1993) and adopted in the Northwest Forest Plan. On the other hand, some ecological functions must be sustained on every piece of ground, especially those related to soils, nutrient cycling, and the interactions between land and water that regulate hydrologic flows and produce clean water.

The biological crisis in the Pacific Northwest can be traced directly to a severe imbalance in land-use types. During the latter half of the 20th century, intensive forest management spread to cover a majority of the regional landscape. With the possible exception of the northern Rockies (Idaho and western Montana), which have a high proportion of area in designated wilderness, the amount of area in reserves was grossly inadequate to sustain old-growth dependent species. Through implementation of plans developed by FEMAT, this imbalance is being redressed in the range of the northern spotted owl. Change is also occurring on federal lands outside the range of the northern spotted owl, especially in eastern Oregon and Washington, albeit more slowly. Throughout the region, but especially outside the range of the northern spotted owl, the degree to which changes are adequate to sustain ecological processes and old-growth dependent species remains to be seen. The interagency scientific analysis of the interior Columbia Basin recommended a management option for federal lands that emphasizes "active" management across the landscape rather than an increased commitment to reserves, arguing that was the best way to ensure restoration and maintenance of ecosystem health (USFS 1996). Managers appear to be left with virtually complete discretion about what form active management might take.

The basic challenges for federal lands throughout the region are to (a) ensure a reasonable probability that the combined area in reserves and forests managed for ecological values is sufficient to maintain ecological functioning and biological diversity, and (b) address the scientific questions necessary to accomplish that successfully. With regard to reserves, the key questions center on how many, how large, where, and the degree of management needed to restore and maintain forest health. With regard to lands managed with an emphasis on ecology rather than production, the key question is what that means in terms of practices. These issues, especially the latter, are not restricted to public lands. With the growing influence of certification of forest products as having come from forests that have been managed in an environmentally suitable manner and general public awareness over environmental issues in forestry, private landowners are increasingly exploring management strategies that balance ecology and economics, a middle ground in which both objectives are served to some degree. In the remainder of this section we briefly discuss current efforts and ongoing issues related to translating the principles of ecosystem management to the ground.

Reserves

Reserves are areas "managed primarily to maintain (or restore) the natural processes and conditions present prior to European settlement" (Aber et al 2000). Reserves typically exclude activities such as timber harvest and road building, but may allow management aimed at restoring or maintaining desired conditions. In some cases, limited timber harvest may be permitted if deemed necessary to lower risk of catastrophic disturbances.

The science of reserve selection and placement has been recently reviewed by Noss and Cooperrider (1994) and Schwartz et al. (2000). Recent models developed specifically for spotted owls include Andersen and Mahato (1995) and Hof and Raphael (1997). It is beyond the scope of this report to deal with this topic in detail, however some specific issues are important to address, particularly the current debate about whether reserves are a wise strategy for maintaining diversity in the long run. The literature on the topic points out that the majority of scientists with expertise in this issue believe adequate and well-distributed reserves are essential components of any conservation strategy (e.g. Noss 1983, Harris 1984, FEMAT 1993, Aber et al 2000).

A contrarian view has been expressed by some forest managers and academic foresters, who argue the best way to protect habitat in the long run is active management, including timber harvest. This view was expressed in greatest detail in the Report on Forest Health in the United States, written by a panel chartered by Congressman Charles Taylor (Oliver et al. 1997), and the approach was recommended by the Interior Columbia Basin Ecosystem Management Project (USFS 1996). Proponents of active timber harvest on all or most of the landscape argue their approach reflects current ecological thinking, which recognizes nature as inherently dynamic, from which it follows that attempts to freeze any particular set of conditions are futile and misguided. Only through active timber harvest can stands be maintained in a healthy condition that continues to supply habitat (Oliver et al. 1997). This view recently was critiqued by a panel of the Ecological Society of America (ESA), which disagreed strongly with the conclusions of Oliver and colleagues (Aber et al. 2000). This committee concurs with the ESA panel, which based its conclusions on three factors:

- Although current ecological thinking indeed recognizes nature as dynamic rather than static, what this implies for conservation is not that reserves are a bad strategy, but that reserve design must account for dynamism in nature, largely by (a) ensuring enough area in reserves to subsume natural disturbance and recovery processes (e.g. FEMAT 1993), and (b) designing reserves so as to reduce risk of catastrophic disturbance spreading from beyond their boundaries, which generally, though not necessarily always, means relatively large reserves well buffered at their boundaries (Noss 1983; Andersen and Mahato 1995; Schwartz et al. 2000).

- There is little evidence that managed stands are healthier than unmanaged stands. In fact, quite the contrary, experience both within and without the Pacific Northwest shows stands managed for timber are more susceptible to crown fires, pathogens, and insects than the natural forests they replaced (Huff et al. 1995; Schowalter et al. 1997; Perry 1998; Aber et al. 2000).
- Even in the eventuality that forests could be logged so as to maintain habitat for the entire suite of old-growth dependent species, reserves still play a vital aesthetic and spiritual role for humans.

A well-designed and properly managed reserve system, in which appropriate levels of management may play a legitimate role is an important feature of forest management.

Logging to Improve Forest Health

A number of forests in the Pacific Northwest that developed after clearcutting (moist forest types) or high-grade logging coupled with fire exclusion (dry forest types) are severely overstocked, a situation that was apparently uncommon before the arrival EuroAmericans (Mutch et al. 1993; Tappeiner et al. 1997). Overstocking increases fire hazard in moist forest types west of the Cascades crest, and creates a variety of problems in dry forest types of the Klamath province and throughout the interior, including increased susceptibility to crown fires, insects, and pathogens (Perry 1988b; Mutch et al. 1993). Where overstocking threatens forest health, some careful logging may be appropriate even within reserves. FEMAT (1993), for example, proposed thinning in overstocked young stands within reserves to reduce fire hazard. Thinning understory trees within old-growth ponderosa pine forests has been proposed to reduce the risk of crown fire and insect infestation (e.g., Mutch et al. 1993), although many biologists and environmentalists are skeptical that benefits outweigh the environmental risks associated with logging.

Logging to improve forest health is a complex issue that eludes simple generalizations. Thinning overly dense stands undoubtedly improves individual tree vigor, which in turn improves tree resistance to bark beetles, especially among the pines. Thinning may or may not reduce fire hazard, depending on which trees are thinned and how logging slash is treated. Thinning that lowers average stand diameter may well increase susceptibility to crown fires (Perry 1995a), and opening the stand too much could dry fuels and increase flammability. Experience from northern California shows partially logged stands in which logging slash is left untreated are more susceptible to crown fires than unlogged stands (Weatherspoon and Skinner 1995). On the other hand, underthinning removes fire ladders that propogate flames into the crown of overstory trees, and hence reduces the chance of a ground fire becoming a crown fire. To successfully achieve the objective of reducing fire hazard, all these factors must be taken into account and protocols developed for deciding which stands should be thinned and how much. Potential negative affects of logging should also be considered. Given the well-documented environmental problems associated with roads (Trombulak and Frissell 2000), logging to improve forest health would not seem justified if it required building roads into roadless areas. Issues of habitat must be considered. Some animals, notably goshawks and ungulates, require closed canopy forests in at least a part of their range and could be negatively impacted by widespread heavy thinning.

In sum, judicious logging may be a good tool to improve the health of some forests, but each situation should be evaluated on its own merit, and operations planned carefully to ensure the cure is not worse than the disease.

Management for Complexity and Diversity

Much of human management is focused on ecosystem "simplification" or techniques to focus productivity into those elements of particular human interest. However, much research in ecology and forest management has shown that biological diversity and structural complexity of ecosystems are critical to such ecosystem processes as primary production and nutrient cycling. Complexity and diversity also impart resistance to and resilience from disturbance and provide the genetic resources necessary to adapt to long-term change.

Three management strategies are key to the conservation of biological diversity. First, management practices for any one species must recognize that suitable habitat encompasses all of the other species and system processes upon which that species depends. Second, area of habitat required to sustain viable species populations must be sufficiently large to buffer inevitable variations in population size through time. Third, as discussed above, a landscape and regional approach to the distribution of reserves and connections among them is critical.

Variability and Change

Forest ecosystems are constantly changing. Natural disturbances such as fire, windstorms, insect and pathogen epidemics, and floods are ubiquitous and, in many cases, critical to the maintenance of key ecosystem processes. Management determined to "freeze" ecosystems in a particular state is futile and unsustainable.

Nowhere is the importance of variability and change better demonstrated than with natural disturbances and the processes of ecological succession that derive from them. As discussed in Chapter 4, that issue is critical in the dry forest types of the Eastside (Mutch et al. 1993; USFS/BLM 1994; Hessburg et al. 1993; Sampson and Adams 1994). Fire hazard is high, outbreaks of defoliating insects and, in some places, bark beetles have been more severe than at any time within the recorded past (e.g. Anderson et al. 1987; Wickman et al. 1992; USFS/BLM 1994), and the stage is set for severe pathogen problems (Hessburg et al. 1994). There is consensus that the current poor state of those forests can be traced in part to decades of fire exclusion, high grade logging of ponderosa pine and western larch, and overgrazing (Anderson et al. 1987; Mutch et al. 1993; Covington et al. 1994; see Chapter 4).

Fire exclusion in many Eastside areas has resulted in large-scale conversion from a landscape dominated by old-growth, largely ponderosa pine, to forests that are younger, much denser, and dominated by tree species susceptible to western spruce budworm,

A Framework for Sustainable Forest Management

Douglas-fir tussock moth, and several root rots. Reduced habitat (e.g., large dead wood and old-growth forests) for the natural enemies of defoliating insects may have exacerbated outbreaks (Torgersen et al. 1990; Bull and Holthausen 1993). The combination of densely stocked young trees and (in many stands) large amounts of dead fuels resulting from insect outbreaks has significantly increased the probability that what in the past would have been a gentle ground fire will now become a severe crown fire.

The following steps will be critical to dealing with this situation: thinning to reduce tree density; reintroduction of frequent ground fires; restricting any harvest in remnant old-growth stands to thinning understory trees; protection of any mature or old-growth ponderosa pine trees; maintenance of coarse woody debris as habitat for natural enemies; and longer rotations to restore the landscape to a high proportion of large, fire-resistant early successional tree species (primarily ponderosa pine) (Perry 1988a, Mutch et al. 1993, Hessburg et al. 1993, Torgersen 1993, Arno et al 1993, Bull and Partridge 1986, Covington and Moore 1994, Henjum et al. 1994, O'Laughlin 1994, Oliver et al.1994, USFS/BLM 1994). However, the areas affected are quite large, and the remedial prescriptions are in some cases very expensive.

Human development and building on fire-prone landscapes has increased the financial liability and risk to human life associated with catastrophic events that might occur as a consequence of extensive fuel accumulations. At the same time, such development severely constrains the variety of management interventions that may be used to remedy the situation. Prescribed fire in heavy fuel areas close to homes or other structures is itself risky and expensive and must be applied on a limited scale. Furthermore, the use of prescribed fire is limited in some places owing to the effects of smoke on air quality near already polluted urban centers.

Uncertainty and Surprise

Three kinds of uncertainty are inherent in the management of forest ecosystems (Hilborn and Mangel 1997, Christensen et al. 1996). First, uncertainty derives from unknowable responses that result from the complex and ever changing character of ecosystems. Examples of such uncertainty include ecosystem responses to climate change, rare events such as volcanic eruptions, or the cumulative effects of multiple environmental changes (e.g., the multiplicity of factors influencing populations of migratory fishes in forested landscapes). The fact that our forest ecosystems are being stressed in ways that may be unique with respect to their evolutionary history increases this kind of uncertainty. Uncertainties of this kind are difficult to eliminate or reduce, but their magnitude and relative importance can be estimated.

Second, uncertainty arises because of limitations in our knowledge base or models of ecological and social systems. Uncertainties of this sort are being reduced by increased research, although extension of such information across scales of time and space represents a significant challenge (Levin 1992).

Poor data quality, sampling bias and analytical errors generate the third category of uncertainty. Managers and decision makers must work with scientists and data managers to determine an acceptable level of decision error.

Public understanding and education on this matter is critical. Just as managers cannot claim perfect knowledge, the public should not expect it. Public understanding of the nature of uncertainty in our science and management, as well as the potential risks from surprises is a necessary prerequisite to informed involvement in decisions affecting the management of natural resources. Institutional barriers to learning can limit our capacity to reduce uncertainty (Lee 1993). For example, management agencies often lack systematic plans for learning, which should include prioritized listings of identified uncertainties, methods for reducing important and tractable uncertainties, procedures for evaluating existing actions, and mechanisms for retaining new knowledge in the memory of the institution (Hilborn and Mangel 1997, Christensen et al. 1996). Dealing with these barriers is critical to the accountability and adaptability of management.

Humans as Ecosystem Components

Humans present some of the most significant challenges to sustainability, but they are also integral ecosystem components and must be considered as such in any effort to achieve sustainable-management goals. Given the growth in human populations, sustainable provision of ecosystem goods and services becomes an even more compelling goal. Humans as ecosystem components should be considered in at least three ways: 1) their use of all resources in the ecosystem, 2) their use of the land, and 3) their effects as a result of nonindustrial private forests owners' decisions and goals.

The principles outlined earlier in this chapter must form the foundation for forest management if the goals and challenges posed in this chapter are to be met over the long term.

Making Management Adaptable

Forest-management practices have long been adapted in response to changes in markets, social values, uncertainty and risk, and available information about the effects of management practices. Perhaps because change in these conditions in the Pacific Northwest is occurring at a more rapid pace than previously, or perhaps because the concepts of spatial and temporal scale of management are expanding, the term "adaptive management" is now being used in the region to describe a management scheme that pays particular attention to uncertainty and the general paucity of information about the effects of management on the sustainability of forests.

Of necessity, natural resource managers set goals and implement practices with an incomplete knowledge base. Given the complexity and variability inherent in the natural world, this will always be the case. Thus, many treatments of ecosystem management have emphasized that management should be viewed as experimental and that good management should include the means to learn from our experiments and adjust goals and practices accordingly. This concept of "management as experiment" is the basis for adaptive management. Walters (1986) proposed the following elements for adaptive management:

• Management practices should be framed by clearly stated objectives and bounded by an honest appraisal of the practical constraints on action.

• Existing understanding of managed systems should be represented in terms of explicit models of dynamic behavior that spell out assumptions and predictions clearly enough so that errors can be detected and used as a basis for further learning.

- Uncertainty and its propagation through time in relation to management actions should be understood using statistical measures and imaginative identification of alternative hypotheses (models) that are consistent with experience but might point toward opportunities for improved management success.
- Policies should be balanced to provide for continuing resource production while simultaneously probing for better understanding and untested opportunity.

Thus, adaptive management explicitly provides a basis for learning as forest management proceeds to meet human-defined goals. It is based on a learning process that distinguishes among what is known, what is suspected, and what is not known. Adaptive management can be viewed as a series of actions that are tentative, that are watched closely, and that are changed as their results become apparent. It should be viewed as complementing, rather than replacing, reliance on formal research results.

Monitoring the results of forest-management actions is needed to identify successful practices and to recognize when changes are necessary. Given the limitations in understanding of the behavior of, for example, managed forest ecosystems, managing without monitoring elements critical to meeting management goals is akin to trying to navigate without a compass (Lee 1993). This is as true for monitoring social and economic factors as it is for monitoring ecosystem attributes.

The scale and intensity of monitoring programs must be consistent with the scales of the processes that are to be monitored (Levin 1992). The Forest Service monitors some attributes of forest conditions at fairly gross geographic and temporal scales in the Forest Inventory and Analysis program (FIA) of its research branch. Although that level of monitoring provides some useful information on changes in forest conditions over periods of decades and at a multicounty spatial scale, it has not been designed to monitor changes in forest conditions at the level of a forest stand or in response to particular management actions.

The scale at which attributes of a managed forest should be monitored needs to be addressed (Levin 1992). Sampling design, technical aspects (e.g., logistics, costs, and equipment), and institutional and policy issues involving responsibility and coordination all need to be considered in monitoring to support effective adaptive management. Because forest ecosystems can buffer environmental changes that can be deleterious

A Framework for Sustainable Forest Management

over time, monitoring to detect early changes present a special challenge. For example, trees and forests can look healthy even after irreversible damage from insects or pathogens. A clear understanding of the processes that underlie forest vitality is critical to the detection of early and often subtle indicators. That is one aspect of monitoring for adaptive management that needs further attention.

Resolving Conflicts

Processes for resolving conflicts in the management of Pacific Northwest forests clearly have not worked well. Conflicts are caused by differences in value systems (e.g., the marketplace versus the political system), requirements of the present versus those of future generations, spatial issues (e.g., issues that cut across ownerships or political boundaries), and the time required for societal institutions (e.g., the legal structure, markets, and agencies) to respond and change.

Yaffee (1994) suggests the need for four improvements in institutions to avoid or manage problems of the type exemplified in the disputes over the northern spotted owl: (1) new mechanisms to bridge the agency-nonagency boundary to build understanding and political concurrence; (2) altered approaches to organizational management, including updated notions of leadership; (3) improved means of gathering and analyzing information about resource problems, organizational possibilities, and political and social context; and (4) ways to promote a culture of creativity and risk-taking to generate more effective options for the future (Yaffee 1994). It is evident from this listing that no single approach or new policy institution will solve the problems faced in managing the forests of the Pacific Northwest.

In the spirit of adaptive management, various efforts in the region are being made to find ways to resolve some of the conflicts over forest management. Some of these are led by stakeholders such as the states, large landowners, or local groups of citizens. For example, under a process fostered by the Washington Forest Practices Board and enforced by the Washington Department of Natural Resources, landowners are cooperating in a program of watershed analyses to resolve issues involving timber, fish, and wildlife at the individual watershed level. Cooperative projects, some of which have been identified for "adaptive management units" under the Northwest Forest Plan, have been initiated by groups of citizens and governmental agencies in several parts of the region. Many articles and reports discuss such cooperative projects, including Wondolleck (1988) and Endicott (1993).

The FEMAT process for federal lands is another approach, one that has been driven by the need to accommodate forest management of federal lands to the requirements of the Endangered Species Act (ESA). Various "habitat conservation plans" (HCPs) have also been developed by owners of private forests in response to the requirement of the ESA that habitat for endangered species cannot be reduced if it will endanger protected wildlife. The specific objective of most of these HCPs is to avoid the "incidental take" provisions of the ESA regulations and thereby to reduce some of the uncertainty in dealing with these regulations. Other approaches include one proposed by Gordon (1994) that an effort be launched to extract a national public vision to underpin and guide forest policy and management.

No single approach is best or, by itself, sufficient. The obvious need is to keep experimenting with ways to resolve conflicts in forest management, especially at the local level, in full recognition of the existing limitations of the overall institutional structure for doing so.

9
CONCLUSIONS AND RECOMMENDATIONS

No single phrase, concept, or picture can capture the breadth of forest management in the Pacific Northwest or the environmental issues that are involved in it. The region is large and geographically varied. Its forests are complex. Although the public policy debates over forest management have focused on federal forests, ownership is varied. There are large and more or less contiguous blocks of state, Indian, and forest industry forests throughout the region. Privately owned forests other than those in forest industry ownership, and often in relatively small tracts, are mingled among the larger ownerships. The conclusions and recommendations that follow in this chapter must necessarily be drawn in terms that give full recognition to the complexities of the region's forests and its forest ownerships.

Management of federal forests in the region has changed dramatically over the past two decades. The Northwest Forest Plan for federal forests in the range of the northern spotted owl west of the Cascades Crest brought the most dramatic and immediate changes. But even it was the continuation of a process of change that has affected management of federal forests throughout the region and, indeed, in other parts of the country as well. Although change in forestry practices has also occurred on other forests, public and private, it has been more gradual. It is clear, however, that what happens on any of the major categories of forest ownership in the region affects practices on the others.

FORESTRY PRACTICES IN THE PACIFIC NORTHWEST

Protected reserves: The chief role of federal forests in the region has shifted from one of providing timber and other forest products to one of sustaining and restoring forest ecosystem integrity. Reserves have been established for various purposes: as elements of the National Wilderness Areas Preservation System, as protected natural areas, and most recently on the Westside to protect the northern spotted owl and other threatened species. Some have been selected because they contain old growth and others because they provide amenities not necessarily associated with old growth. The common characteristic is their relatively natural conditions. Management of the matrix of federal forests that encompasses the reserves has shifted from an emphasis on timber production to an emphasis on sustaining ecological conditions.

Intermingled with the federal forests are nonfederal forests managed by a diverse set of owners with a wide range of goals. Most forest industry land is managed intensively for timber. The rest of the nonfederal forests is managed at various intensities to meet goals that range from timber production to full protection of natural values.

This not wholly coherent amalgam of more-or-less natural reserves scattered across a forested landscape of tracts, large and small, managed for various purposes reflects the shifting policies for federal forests as well as the changing management practices on nonfederal forests. To suggest that there should be a single coherent policy framework for all Pacific Northwest forests is probably futile. To suggest that even the policy framework for the region's federal forests should be stable and the same across the entire region is probably unwarranted. But the committee believes some elements of such a policy framework can be defined.

The committee believes that forest-management goals should include conservation and protection from harvest of the vast majority of late-successional and old-growth forests in the Pacific Northwest. The long-term values of conserving remaining late-successional and old-growth forests in the Pacific Northwest are great. Further cutting of the remaining late-successional and old-growth forests will accelerate threats to the biological diversity of the Pacific Northwest and threaten our ability to sustain important ecosystem processes. Protected reserves

Conclusions and Recommendations

have a role in meeting these goals, especially in the absence of other accepted mechanisms for meeting them.

Recommendation: Forest management in the Pacific Northwest should include the conservation and protection of most or all of the remaining late-successional and old-growth forests. Protected areas that include late-successional and old-growth forests should have an important role in an overall strategy for forest management in the region.

The reserves established by the Northwest Forest Plan are a step in the right direction. Meeting the broader goals of sustaining landscape level natural processes and maintaining the region's biodiversity legacy will, however, require further attention to the role of reserves in general, the character of these reserves in particular, and the suitability and availability of other approaches. This is particularly relevant in the parts of the region outside the range of the northern spotted owl.

Not all ecosystem types are represented on public lands in the Pacific Northwest–examples of gaps in coverage include lowland floodplain forests, oak woodlands, and coastal tidal marshes. Checkerboard ownership of public and private lands hinders effective management of forest ecosystem patterns and processes. Opportunities such as land exchanges might offer ways to obtain critical habitats and create public and private management boundaries that are consistent with the behavior of ecosystem processes.

Recommendation: Goals for protected late-successional and old-growth reserves should include representation of the range of forested ecosystems in the region. This should include rationalization of reserve boundaries, and land exchanges between public and private landowners should be pursued.

Other forests: All forests, federal and nonfederal, need to be managed at appropriate scales of time and space. In the Pacific Northwest, where the forests have been and will continue to be subject to strong and varied human pressures, management at the landscape level must consider the roles of both reserves and managed forests (Perry 1998). This necessarily involves consideration of forest structure and functioning at the level of the stand, the spatial patterns of stand structures at the landscape level, and the temporal dynamics of both stand and landscape structure that result from disturbances.

The effects on ecological processes and biological diversity of natural disturbances and those caused by human activities are highly variable. Variations in the scale at which disturbances occur are poorly understood, but they are known to be significant. The inevitability of catastrophic disturbances that may reset successional processes should be acknowledged in the design and extent of late-successional and old-growth preserves. For managed forests, silvicultural treatments intended to simulate the effects of disturbances must be designed with attention to specific goals. These include effects on fuel loads, structural features, or legacies important to postdisturbance regeneration, biodiversity, patterns of postharvest recovery, and landscape flammability.

Natural disturbances such as fire and severe wind are an integral part of Pacific Northwest forests, although the spatial and temporal patterns of such disturbances vary among ecosystem types. Alteration of natural disturbance cycles has had adverse effects on the condition and diversity of some forested landscapes in the region. Extensive accumulations of fuel have put Eastside forests and landscapes at risk from intense and extensive wildfires. In many Westside forests, losses of tree diversity and structural complexity that normally result from snags and downed logs have altered ecosystem processes such as nutrient cycling and water movement and might have increased risk from epidemics of insects and disease.

The legacies of disturbance—residual woody debris, ash accumulations, seed banks, advanced regeneration, and surviving organisms—greatly influence change after disturbance (Franklin 1993b). The legacies vary among fire regimes and between natural disturbances and management interventions. Managers cannot control these phenomena, but they can and should protect and, where necessary, restore the natural mechanisms by which ecosystems are buffered from such change.

Recommendation: The important roles of natural disturbances and legacies in sustaining ecological processes must be recognized in forest-management practices for both federal and nonfederal forests in the Pacific Northwest.

Conclusions and Recommendations 203

Given the spatial and temporal variability of disturbance processes, collaborations across boundaries of ownership and jurisdiction will be necessary. Natural disturbances occur at spatial scales that transcend ownership and political boundaries. Fires can burn across vast areas, influenced by regionwide variations in fuels and climate; population processes that ensure viability of species, especially wide-ranging vertebrates, have little to do with the borders of jurisdictions; and the functioning of aquatic ecosystems depends on the continuity and integrity of entire hydrologic systems well beyond the reach of an individual stream.

A formalized approach to adaptive management is needed to evaluate effects of new forestry practices on key ecosystem properties and to adjust management practices in a timely fashion to changes in forest condition across all spatial scales. The approach should include elements such as acknowledgment of ignorance, identification of key ecosystem processes, model development, monitoring that focuses on management objectives, standards for data evaluation and data quality, and timely feedback.

For example, the role of pathogens and insects in sustaining Pacific Northwest forests is clearly important, but is poorly understood. At the same time, outbreaks of both are becoming increasingly severe and widespread. The place of various methods for controlling outbreaks and their potential effect on sustaining ecological processes are also poorly understood. Adaptive management approaches for dealing with such outbreaks could be used to recognize that decisions must be made in the absence of solid information.

Recommendation: A formalized approach for adaptive management should be developed and applied in evaluating the effects of forest management practices on key ecosystem properties and to guide changes in these practices that reflect forest conditions at all spatial scales.

Our understanding of the full costs of using forest resources is also changing as pressure on these resources to meet human demands mounts and as scientific knowledge grows. In some cases, recognition of the full costs comes only after use has occurred. Examples include decreases in the ability of forest land to sustain long-term timber production as a result of practices that significantly lower the productivity of the forest or that cause offsite impacts and the loss of species that are dependent on old-growth forests.

Management strategies to restore ecosystem features will be important in a long-term conservation plan. Such management might include reduction of flammable fuels in Eastside forests or enhancement of structural complexity in Westside forests. Forest managers will need to develop and implement a combination of protocols using of prescribed fire, as well as silvicultural and harvest techniques, to reduce accumulated fuels where such accumulations increase risk of catastrophic wildfire. Those protocols should be sensitive to protection of streams, soils, and other vulnerable components of the forest environment.

Economic consequences: The impacts of changing forestry practices on federal lands on overall employment and regional income in the Pacific Northwest have been relatively small. Even the establishment of old-growth reserves on a substantial part of the Westside federal forests has had only minor effects on the region's economy as a whole. At the same time, some communities that had been heavily dependent on federal timber harvests have had a difficult time. Even there, the impacts have been ameliorated by growth of the overall region's economy, as well as by assistance provided by the federal government as part of the Northwest Forest Plan (Tuchmann et al. 1996).

Over the past 20 years in the Pacific Northwest, rural communities have generally become much less dependent on timber, mining, fishing, or agriculture (Anderson and Olson 1991). The region's local economies increasingly benefit from a mix of extractive industries, light manufacturing, retirement, residential, service, and recreation sectors. In Oregon, for example, the lumber and wood products industry now represents only 5% of total employment. Mills have been closed and more jobs have been lost to increases in efficiency and productivity than to reductions in timber harvests. National and world economic forces have had a more direct bearing on employment than has timber supply (Waggener 1990).

Before adoption of the Northwest Forests Plan, disputes over likely economic effects of reductions in federal timber harvests were heated. The range of employment losses projected by opposing interest groups was broad in large part because estimating the regional economic effect

Conclusions and Recommendations

of shifts in the proportions of wood and nonwood products resulting from changes in timber harvests is difficult (Sample and Le Master 1992). This suggests the need to improve the analytical and information bases for relating changes in forestry policies at various levels to possible economic consequences.

WHAT IS OLD-GROWTH?

Old-growth forests are defined as those that have accumulated specific characteristics related to tree size, canopy structure, dead snags and woody debris, and an assemblage of particular plant and animal species that inhabit them. These specific attributes of old-growth forests develop through the process of forest succession until the collective properties of an old-growth forest are evident. No simple measure can be used to define an old-growth or late-successional forest.

The ecological characteristics and appearance of old-growth forests vary among forest types across the Pacific Northwest. Increasingly, definitions rely on indexes of successional development based on multiple forest characteristics. Current definitions used by the U. S. Forest Service use specific values or states for five criteria – number of large old trees per acre, variation in tree diameters, degree of tree decadence, amounts of large dead wood, and characteristics of the canopy structure.

While many of the characteristics that define old-growth forests develop during the second century of stand development, all of the properties of an old-growth forest typically are not present until the forest is at least 200 years old. The committee emphasizes that it is the presence of the assemblage of characteristics that determines whether a forest can be classed as old-growth and not a specific age. The committee also emphasizes that the defining assemblage of characteristics will vary in forests across the Pacific Northwest.

An old-growth forest in good condition is one that retains its basic structures and processes (Rapport 1989). Old-growth forests are biotically more complex than forests in earlier successional stages. Compared with younger forests, they have a greater diversity of ecosystem components and specialized organisms and produce more food for some animal species. They have a higher total amount of live and dead biomass and a higher amount of woody debris in streams and terrestrial areas. Old-growth forests are also less susceptible to large-scale disturbances and pest outbreaks, and they have a lower incidence of root-rot problems. They have unique microclimates and might have an effect on regional climate as well.

OLD-GROWTH MANAGEMENT

Management in the form of establishing reserves is clearly an important tool in maintaining existing old-growth forests in the Pacific Northwest. Other kinds of management interventions in natural processes are also appropriate to some extent in such reserves. For example, protection against destructive wildfires may be advisable, although the role of fire and other disturbances in sustaining natural processes in the reserves must also be recognized. Other kinds of management practices can also have a role within and outside of reserves in encouraging the development of some of the properties of old-growth forests. In such cases, it must be recognized that the defining characteristic of an old-growth forest is the assemblage of various properties that goes beyond just age of trees and canopy structure.

Managed forests can be thinned to produce large trees and structural heterogeneity at a relatively early age, especially in areas of high site quality. But accelerating the development of these two conditions does not by itself result in old-growth forests. Whether this accelerated development successfully mimics processes that can produce other old-growth properties is unclear at the present time. Management approaches such as "green tree retention," which attempt to mimic natural disturbances by preserving decaying logs and soil organic matter, with time might lead to managed forests with at least some old-growth characteristics (McComb et al. 1993). The degree to which this will occur can be determined only with extensive testing across a range of forest types and conditions.

FOREST PRODUCTS SUBSTITUTION

Recent reductions in federal timber harvests in the Pacific Northwest have been met by increased timber harvests in the South and increased softwood lumber imports from Canada. Together, increased production in the South and increased imports from Canada, both in response to ordinary market forces, have offset the reduced softwood harvests on federal forests in the West. Substantial substitution for Pacific Northwest wood products from the southern hemisphere or from Europe and temperate parts of Asia has not occurred so far. Total consumption of softwood wood products in the United States does not appear to have been substantially reduced. The reduction in federal timber harvests has been accompanied by some increase in the price of softwood lumber to consumers and in the prices paid for timber that is harvested from both federal and nonfederal forests.

The expected effects of adopting the Northwest Forest Plan on some biological resources in the Pacific Northwest have been examined at length (FEMAT 1993). The potential effects on most other nonwood products, such as recreation and special forest products, were not thoroughly evaluated in that report, partly because of the lack of good information. But it is clear that the reductions in federal timber harvests in the Pacific Northwest favor some kinds of both game and nongame species of wildlife over others, affect hunting conditions, improve habitat for fisheries, and maintain opportunities for some kinds of recreation in the region.

The extent to which these impacts will affect interregional markets for these and other nonwood forest products is unclear. Most nonwood forest products are sold in local and regional markets, although some, such as wild-grown mushrooms, may end up being used far from their source. Although solid information is lacking, it does not seem likely that changes in the availability of nonwood forest products in the Pacific Northwest will be reflected in markets for these or competitive nonwood products in other forest regions in the United States.

Sustaining the increased level of timber harvests in the South, which come mainly from private forests, will require more intensive management practices. The possible effects on biological resources, such as wetlands and the red cockaded woodpecker, of more intensive practices brought about by the decrease in Pacific Northwest timber harvests have apparently not been carefully evaluated. Similarly, possible effects on employment and communities in the South have not been carefully evaluated.

Pressures on forests for all uses in the Pacific Northwest and elsewhere in the United States will probably continue to rise. The specific demands on forests may change, but the basic demands for materials, space, and environmental amenities will almost certainly continue to increase. The increasing production of timber on private forests is leading to lower ages of trees at harvest and more intensive silvicultural operations such as thinning, use of improved genetic stock for single-species planting, fertilization, and the increased use of herbicides and insecticides. Tracking these changes and being prepared to take action when the effects are judged to be serious are challenges for public policy.

Recommendation: Regional assessments of the impacts of increasingly intensive forest management practices, especially on private forests, should be conducted to evaluate the impacts of shifting regional patterns of timber harvesting. In particular, an assessment is needed of the effects on key species and ecosystems in the U.S. South of increased timber harvests and management intensity that has resulted from reduced timber harvests on federal forests in the West.

Existing institutional structures do not appear to have been adequate for planning and managing forest management activities over a variety of spatial and temporal scales. Few mechanisms are available to facilitate dialogue and resolve conflicts at large spatial scales involving multiple ownerships and many stakeholders. The committee believes empirically based, comparative evaluations of the effectiveness of alternative strategies are needed. Federal agency and constituency dialogue should be reviewed with particular attention to procedures and policies through which public agencies can communicate with stakeholders.

Recommendation: Experience with FEMAT, the Northwest Forest Plan, and other processes used to help resolve disputes over Pacific Northwest forestry practices should be used to explore alternative mechanisms for dispute resolution.

RESEARCH RECOMMENDATIONS

Limitations on available knowledge for guiding forest management and

Conclusions and Recommendations

resolving issues in the Pacific Northwest have been noted throughout this report. An accelerated program of research is needed to fill these gaps. Parallel gaps in knowledge exist for other regions of the country as well. The committee's conclusions support in general terms the significant reorientation of forestry research recommended by an NRC panel in 1990 (NRC 1990).

Various institutions and sources of funding play important roles in forest-related research in this region and in the country as a whole. The federal role in funding both in-house and extramural research is obviously very important, but the states, forest industry, and nonprofit organizations also provide research support. The committee believes all of these institutions can take part in supporting and conducting the needed research. In particular, the federal government should substantially strengthen its support for a competitive research grants program that would recognize the broad array of scientific specialties and research organizations that are relevant to current issues involving forest management and conservation.

Specific areas of research in need of increased funding and attention include the following:

- the relationship of natural disturbances to the sustainability of protected and managed Pacific Northwest forests and the extent to which the effects of these disturbances can be simulated by management practices;
- the relative importance of legacies and their role in maintaining forests and regenerating harvested areas, and the extent to which management actions can "create" legacies;
- the role of insects and pathogens in sustaining natural processes in Pacific Northwest forests and factors involved in insect and pathogen outbreaks in the region;
- forest restoration methods and their role in restoring and maintaining forest vitality;
- the impacts of forest-management practices, including timber harvesting, on the production of nonwood forest products, including recreation and special forest products such as wild-grown mushrooms;
- information for making accurate assessments of the impacts of changes in forest practices on regional and local employment and income;

- the impacts of changes in forest practices in the Pacific Northwest on biological and nonbiological factors within the region and in other affected regions;
- continued basic research on the biological functioning and interactions of the multitude of life forms present in the Pacific Northwest forests.

REFERENCES

Aber, J., N. Christensen, I. Fernandez, J. Franklin, L. Hidinger, M. Hunter, J. MacMahon, D. Mladenoff, J. Pastor, D. Perry, R. Slangen and H. van Miegroet. 2000. Applying Ecological Principles to Management of the U.S. National Forests. Issues in Ecology 6:1-20.

Adams, D.M. and R.W. Haynes. 1980. The 1980 Softwood Timber Assessment Market Model: Structure, projections, and policy simulations. Forest Science Monograph No. 22, Society of American Foresters, Washington, DC. 64 pp.

Adams, D.M., K.C. Jackson, and R.W. Haynes. 1988. Production, Consumption, and Prices of Softwood Products in North America: Regional Time Series Data, 1950 to 1985 (later extended to 1993). Resource Bull. PNW-RB-151. Portland OR.: USDA, Forest Service, Pacific Northwest Research Station. 49 pp.

Adams, D.M., R.J. Alig, B.A. McCarl, J.M. Callaway, and S. Winnett. 1996. An analysis of the impacts of public timber harvest policies on private forest management in the U.S. Forest Science 42(3):343-358.

Agee, J.K. 1981. Fire effects on Pacific Northwest forests: flora, fuels, and fauna. Northwest Fire Council Proc. 54-66.

Agee, J.K. 1991. Fire history along an elevational gradient in the Siskiyou Mountains, Oregon. Northwest Science. 65(4):188-199.

Agee, J.K. 1993. Fire Ecology of Pacific Northwest Forests. Washington, D.C.: Island Press.

Aizen, M.A. and P. Feinsinger. 1994. Habitat fragmentation, native insect pollinators, and feral honey bees in Argentine "Chaco Serrano." Ecological Applications 4(2):378-392.

Akçakaya, H.R., M.A. Burgman, and L.R. Ginzburg. 1999. Applied Population Ecology : Principles and Computer Exercises Using RAMAS EcoLab 2.0, 2nd Ed. Sunderland, Mass. : Sinauer Associates.

Amaranthus, M.P., J.M. Trappe, L. Bednar, and D. Arthur. 1994. Hypogeous fungal production in mature Douglas-fir forest fragments and surrounding plantations and its relation to coarse woody debris and animal mycophagy. Can. J. For. Res. 24(11): 2157-2165.

Andersen, M.C. and D. Mahato. 1995. Demographic models and reserve designs for the California spotted owl. Ecological Applications 5(3):639-647.

Anderson, H.M. and J.T. Olson. 1991. Federal Forests and the Economic Base of the Pacific Northwest. Washington, D.C.: The Wilderness Society.

Anderson, L., C.E. Carlson and R.H. Wakimoto. 1987. Forest fire frequency and western spruce budworm outbreaks in western Montana. For. Ecol. Manage. 22(3-4):251-260.

Anderson, R.L. 1990. Effects of global climate change on tree survival and forest pest in the South. Pp. 176-180 in: Proceedings of the Society of American Foresters National Convention. Paper presented at the meeting on, "Are Forests the Answer", held July 29-Aug 1, 1990, Washington DC. Bethesda, MD: The Society.

Andrews, H.J. and R.W. Cowlin. 1940. Forest Resources of the Douglas-fir Region. Misc. Publ. 389. Washington, DC: U.S. Department of Agriculture, Forest Service. 169pp.

Anthony, R.G., M.G. Garrett and C.A. Schuler. 1993. Environmental contaminants in bald eagles in the Columbia River Estuary. J. Wildl. Manage. 57(1):10-19.

Arno, S.F. 1980. Forest fire history in the northern Rockies. J. For. 78(8):460-465.

Arno, S.F., E.D. Reinhardt, and J.H. Scott. 1993. Forest Structure and Landscape Patterns in the Subalpine Pine Type: A Procedure for Quantifying Past and Present Conditions. Gen. Tech. Report INT 294. Ogden, UT: USDA, Forest Service, Intermountain Research Station.

Arnolds, E. 1991. Decline of ectomycorrhizal fungi in Europe. Agric. Ecosyst. Environ. 35(2/3):209-244.

Barett, R. 1987. Tourism employment in Montana: quality versus quantity? Western Wildlands 13(2):18-21.

References

Bawa, K.S. 1990. Plant-pollinator interactions in tropical rain forests. Annu. Rev. Ecol. Syst. 21: 399-422.

Beale, C. 1993. Pp. 22-27 in: Poverty is Persistent in Some Rural Areas. Agricultural Outlook. U.S. Department of Agriculture. September 1993.

Becerra, J.X. 1994. Squirt-gun defense in Bursera and the chrysomelid counterploy. Ecology 75(7):1991-1996.

Bechtold, W.A., W.H. Hoffard, and R.L. Anderson. 1992. Summary Report: Forest Health Monitoring in the South, 1991. Gen. Tech. Report SE-81. Asheville, N.C.:USDA, Forest Service, Southeastern Forest Experimental Station.

Beuter, J.H. 1990. Social and Economic Impact of the Spotted Owl Conservation Strategy. Tech. Bull. No. 9003. Washington, D.C.: American Forest Resource Alliance. 37pp.

Beuter, J.H., K.N. Johnson, and H.L. Scheurman. 1976. Timber for Oregon's Tomorrow--An Analysis of Reasonably Possible Occurrences. Oregon State Univ. For. Res. Lab. Res. Bull. 19. Corvallis, OR. 111 pp.

Blahna, D.J. 1990. Social basis for resource conflicts in areas of reverse migration. Pp. 159-178 in: Community and Forestry: Continuities in the Sociology of Natural Resources, R.G. Lee, D.R. Field, and W.R. Burch Jr., eds. Boulder, CO: Westview.

Bliss, J.C., C. Bailey, G.R. Howze, and L. Teeter. 1992. Timber Dependency in the American South. SCFER Working Paper No. 74. Southeastern Center for Forest Economics Research, Research Triangle Park, NC. 15 pp.

Bolsinger, C.L. and J.M. Berger. 1975. The Timber Resources of the Blue Mountain Area, Oregon. USDA Forest Service Resource Bulletin PNW 57. U.S. Forest Service. Pacific Northwest Forest and Range Experiment Station, Portland, OR

Bolsinger, C.L. and K.L. Waddell. 1993. Area Of Old-Growth Forests in California, Oregon, and Washington. Resource Bulletin PNW-RB-197. USDA Forest Service, Pacific Northwest Research Station. December.

Bonnicksen, T.M. 1993. An Analysis of a Plan to Maintain Old-Growth Forest Ecosystems. A report to the American Forest & Paper Association, Washington, D.C. Department of Forest Science, Texas A& M University, Texas.

Booth, D.E. 1991. Estimating prelogging old-growth in the Pacific Northwest. J. For. 89(10):25-29.
Booth, D.E. 1994. Valuing Nature: The Decline and Preservation of Old-Growth Forests. Lanham, MD: Rowman and Littlefield.
Bork, J. 1985. Fire History in Three Vegetation Types on the East Side of the Oregon Cascades. Ph.D. Dissertation. Oregon State University, Corvallis, OR. 94pp.
Bormann, F.E. and G.E. Likens. 1979. Pattern and Process in a Forested Ecosystem. New York: Springer-Verlag.
Boyle, J.R., J.E. Warila, R.L. Beschta, M. Reiter, C.C. Chambers, W.P. Gibson, S.V. Gregory, J. Grizzel, J.C. Hagar, J.L. Li, W.C. McComb, T.W. Parzybok, and G. Taylor. 1997. Cumulative effects of forestry practices: an example framework for evaluation from Oregon, Biomass and Bioenergy 13(4/5):223-245.
Brooks, D., H. Pajuoja, T.J. Peck, B. Solberg, and P.A. Wardle. 1996. Long-term trends and prospects in world supply and demand for wood. Pp. 75-106 in: Long-term Trends and Prospects in World Supply and Demand for Wood and Implications for Sustainable Forest Management. B. Solberg, ed. Research Report 6. Joensuu, Finland: European Forest Institute.
Brown, J.H. and A.C. Gibson. 1983. Biogeography. St.Louis, Missouri: Mosby. 644pp.
Brown, R.B. 1993. Rural community satisfaction and attachment in mass consumer society. Rural Sociology 58(3):387-403.
Brown, T.C. 1999. Past and Future Freshwater Use in the United States: A technical document supporting the 2000 USDA Forest Service Assessment. Gen. Tech. Rep. RMRS-GTR-39. Fort Collins, CO: USDA, Forest Service. 47pp.
Brubaker, L.B., S. Vega-Gonzalez, E.D. Ford, C.A. Ribic, C.J. Earle and G. Segura. 1992. Old-growth Douglas-fir in western Washington. Ecological Studies: analysis and synthesis. 97: 333-364.
Brundrett, M. 1991. Mycorrhizas in natural ecosystems. Adv. Ecol. Res. 21:171-313.
Buchanan, J.B., L.L. Irwin and E.L. McCutchen. 1995. Within-stand nest site selection by spotted owls in the eastern Washington Cascades. J. Wildl. Manage. 59(2):301-310.
Bull, E.L. and R.E. Holthausen. 1993. Habitat use and management of

pileated woodpeckers in northeastern Oregon. J. Wildl. Manage. 57(2):335-345.

Bull, E.L. and A.D. Partridge. 1986. Methods of killing trees for use by cavity nesters. Wildlife Society Bullitin. 14(2):142-146.

Bull, E.L., R.S. Holyhausen and M.G. Henjum. 1992. Roost trees used by pileated woodpeckers in northeastern Oregon. J. Wildl. Manage. 56(4):786-793.

Campbell, R.W. and T.R. Torgersen. 1982. Some effects of predaceous ants on western spruce budworm pupae in north central Washington. Environ. Entomol. 11(1):111-114.

Canadian Council of Forest Ministers. 1997. Compendium of Canadian Forestry Statistics, 1996. National Forestry Database Program. Ottawa: Canadian Council of Forest Ministers. 234 pp.

Cardellichio, P.A., Y.C. Yuon, C.S. Binkley, J.R. Vincent, and D.M. Adams. 1988. An Economic Analysis of Short-Run Timber Supply Around the Globe. CINTRAFOR Working Paper 18, Seattle, WA.: University of Washington. 153 pp.

Carey, A.B. 1995. Sciurids in Pacific Northwest managed and old-growth forests. Ecological Applications 5(3):648-661.

Carey, A.B., J. Kershner, B. Biswell, and L. Dominquez de Toledo. 1999. Ecological Scale and Forest Development Squirrels, Dietary Fungi, and Vascular Plants in Managed and Unmanaged Forests. Wildlife Monographs. No. 142. Bethesda, MD: The Wildlife Society. 71 pp.

Carlson, C.E. and J.E. Lotan. 1988. Using Stand Culture Techniques Against Defoliating Insects. Pp. 275-277. Gen.Tech. Report INT-243. Ogden, UT.: USDA, Forest Service, Intermountain Research Station.

Carter, C. 1988. Assessment of Oregon's forest resource economy: fish and wildlife. Pp. 153-158 in: Assessment of Oregon's Forests. Salem, OR.:Oregon State Department of Forestry.

Carter, M.F. and K. Barker. 1993. An interactive database for setting conservation priorities for western neotropical migrants. Pp. 120-144 in: Status and Management of Neotropical Migratory Birds, D.M. Finch and P.W. Stangel, eds. Gen. Tech. Rep. RM-229. Fort Collins, CO.: USDA, Forest Service, Rocky Mountain Forest and Range Experiment Station.

Castle, E.N. 1993. Rural diversity and American asset. Annals AAPSS 529 (Sept.):12-21.

Chapman, H.H. 1931. Forest Management. Albany, NY: J.B. Lyon. 544pp.

Chen, J., J.F. Franklin and T.A. Spies. 1992. Vegetation response to edge environments in old-growth Douglas-fir forests. Ecological Applications. 2(4):387-396.

Chen, J., J.F. Franklin and T.A. Spies. 1993. An empirical model for predicting diurnal air-temperature gradients from edge into old-growth Douglas-fir forest. Ecol. Modell. 67:179-198.

Christensen, N.L. 1985. Schrubland fire regimes and their evolutionary consequences. Pp. 85-100 in: The Ecology of Natural Disturbance and Patch Dynamics. S.T.A. Pickett and P.S. White, eds. New York: Academic Press.

Christensen, N.L. 1988. Succession and natural disturbance: Paradigms, problems, and preservation of natural ecosystems. Pp. 62-86 in: Ecosystem Management for Parks and Wildernesses, J.K. Agee and D.R. Johnson, eds. Seattle, WA: University of Washington Press.

Christensen, N.L., J.K. Agee, P.F. Brussard, J. Hughes, D.H. Knight, G.W. Minshall, J.M. Peek, S.J. Pyne, F.J. Swanson, S. Wells, J.W. Thomas, S.E. Williams and H.A. Wright. 1989. Interpreting the Yellowstone fires of 1988. BioScience. 39(10):678-685.

Christensen, N.L., A.M. Bartuska, J.H. Brown, S. Carpenter, C. D'Antonio, R. Francis, J.F. Franklin, J.A. MacMahon, R.F. Noss and D.J. Parsons. 1996. The report of the Ecological Society of America committee on the scientific basis for ecosystem. Ecological Applications 6(3):665-691.

Cissel, J.H., F.J. Swanson, G.E Grant, D.H. Olson, S.V. Gregory, S.L. Garman, L.R. Ashkenas, M.G. Hunter, J.A. Kertis, J.H. Mayo, M.D. McSwain, S.G. Swetland, K.A. Swindle, and D.O. Wallin. 1998. A Landscape Plan Based on Historical Fire Regimes for a Managed Forest Ecosystem: The Augusta Creek Study. Gen. Tech. Rep. PNW-GTR-422. Portland, OR: USDA, Forest Service, Pacific Northwest Forest and Range Experiment Station.

Clawson, M. 1976. The national forests. Science. 191(4227):762-767.

Clawson, M. 1979. Forest in the long sweep of American history. Science 204(4398):1168-1174.

Clemens, F.E. 1916. Plant Succession: An Analysis of the Development of Vegetation. Pub. 242. Washington, D.C.: Carnegie Institution of Washington. 512pp.

References

Clemens, F.E. 1928. Plant Succession and Indicators. New York: Wilson. 453pp.

Cohen, W.B., T.A. Spies and G.A. Bradshaw. 1990. Semivariograms of digital imagery for analysis of conifer conopy structure. Remote Sens. Environ. 34(3):167-178.

Colclasure P, J. Moen, and C.L. Bolsinger. 1986. Timber Resource Statistics for the Northern Interior Resource Area of California. Resource Bulletin PNW-RB-135. Portland, OR: USDA, Forest Service, Pacific Northwest Forest and Range Experiment Station.

Committee of Scientists Report. 1999. Sustaining the People's Lands. Recommendations for Stewardship of the National Forests and Grasslands into the Next Century. Washington, DC.: USDA. 193pp.

Connell, J.H. and R.D. Slatyer. 1977. Mechanisms of succession in natural communities and their role in community stability and organization. Am. Nat. 111(982):1119-1144.

Cook, R.E. 1969. Variation in species density of North American birds. Syst. Zool. 18:63-84.

Cordray, S. and K. Goetz. 1994. Comparison of rural and urban mills in a Pacific Northwest county. Annual Meeting of the Rural Sociological Society, August 11-14, Portland, OR.

Courtney, S.P. 1985. Apparency in coevolving relationships. Oikos 44(1):91-98.

Covington, W.W. and M.M. Moore. 1994. Postsettlement changes in natural fire regimes and forest structure: ecological restoration of old-growth ponderosa pine forests. Journal of Sustainable Forestry. 2(1/2):153-181.

Covington, W.W., R.L. Everett, R. Steele, L.L. Irwin, T.A. Daer and A.N.D. Auclair. 1994. Historical and anticipated changes in forest ecosystems of the inland west of the United States. Journal of Sustainable Forestry. 2(1/2):13-63.

Cowles, H.C. 1910. The Fundamental Causes of Succession Among Plant Associations. Report of the Seventy-ninth Meeting of the British Association for the Advancement of Science 1909:668-670.

Cowlin, R.W., P.A. Briegleb and F.L. Moravets. 1942. Forest Resources of the Ponderosa Pine Region of Washington and Oregon. Misc. Publ. 490. Washington, D.C.: USDA, Forest Service. 99 pp.

Crane, M.F., J.R. Habeck, and W.C. Fischer. 1983. Early Postfire Revegetation in Western Montana Douglas-fir Forest. Res. Pap. INT-

319. Ogden, UT: USDA, Forest Service, Intermountain Research Station. 29 pp.

Cromartie, J. 1994. Statement: Before the House Committee on Natural Resources Hearing on "The Changing Needs of the West" Salt Lake City Utah, April 7, 1994. Pp. 81-108 in: The Changing Needs of the West, Oversight Hearing before Committee on Natural Resources House of Representatives, 103 Congress, Second Session, Serial No. 103-80. Washington, D.C.: GAO.

Cubbage, F.W., T.G. Harris, Jr., D.N. Wear, R.C. Abt, and G. Pacheco. 1995. Timber supply in the South: Where is all the wood? J. For. 93(7):16-20.

Culotta, E. 1994. Ecologists gather for mix of Policy, Science in Nashville. Science 265(5176):1178-1179.

Darr, D.R. 1989. RPA Assessment of the Forest and Rangeland Situation in the United States, 1989. Washington DC: USDA, Forest Service.

Davis, D.D., M.L. Torsello and J.R. McClenahen. 1997. Influence of Cryphonectria parasitica basek cankers on radial growth of scarlet oak in Pennsylvania. Plant Dis. 81(4):369-373.

Delcourt, H.R. and P.A. Delcourt. 1991. Quaternary Ecology: A Paleoecological Perspective. New York: Chapman & Hall.

Diamond, J.M. 1972. Biogeographic kinetics: estimation of relaxation times for avifaunas of Southwest Pacific Islands. Proc. Natl. Acad. Sci. USA. 69(11):3199-3203.

Diamond, H.L. and P.F. Noonan. 1996. Land Use in America. Washington, DC: Island Press. 351pp.

Doak, D. 1989. Spotted owls and old growth logging in the Pacific Northwest. Conserv. Biol. 3(4):389-396.

Drielsma, J.H. 1984. The Influence of Forest-Based Industries on Rural Communities. Yale University Ph.D. Dissertation. Ann Arbor: University Dissertation Services.

Drury, W.H. and I.C.T. Nisbet. 1973. Succession. Journal of the Arnold Arboretum 54:331-368.

Edwards, J.S. 1987. Arthropods of alpine aeolian ecosystems. Annu. Rev. Entomol. 32:163-179.

Egan, T. 1994. Oregon, foiling forecaster, thrives as it protects owls. New York Times. Oct. 11:A1, C20.

Egler, F.E. 1977. The Nature of Vegetation, Its Management and

References

Mismanagement: An Introduction to Vegetation Science. Norfolk, Conn.: Egler.

Ehrlich, P.R. and A.H. Ehrlich. 1981. Extinction. The Causes and Consequences of the Disappearance of Species. New York: Random House.

Ellefson, P.V., A.T. Cheng, and R.S. Moulton. 1995. Regulation of Private Forest Practices by State Governments. Minnesota Agricultural Experimental Station Bulletin 605-1995. University of Minnesota, St. Paul, MN. 225pp.

Elo, I.T. and C.L. Beale. 1984. Natural Resources and Rural Poverty: An Overview. Washington D.C.: Resources for the Future.

Elton, C.S. 1958. The Ecology of Invasions by Animals and Plants. London: Methuen &Co. 181pp.

Endicott, E. 1993. Land Conservation Through Public/Private Partnerships. Washington, D.C.: Island Press.

Ewel, J.J. 1986. Designing agricultural ecosystems for the humid tropics. Annu. Rev. Ecol. Syst. 17:245-271.

Ewel, J.J., M.J. Mazzarino and C.W. Berish. 1991. Tropical soil fertility changes under monocultures and successional communities of different structure. Ecological Applications 1(3):289-302.

Fahnestock, G.R. and J.K. Agee. 1983. Biomass consumption and smoke production by prehistoric and modern forest fires in western Washington. J. For. 81(10):653-657.

Farnum, P., R. Timmis and J.L. Kulp. 1983. Biotechnology of forest yield. Science 219(4585):694-702.

FEMAT (Forest Ecosystem Management Assessment Team). 1993. Forest Ecosystem Management: An Ecological, Economic, and Social assessment. Report of the Forest Ecosystem Management Assessment Team. U.S. Department of Agriculture Forest Service, U.S. Department of Interior Fish and Wildlife Service, U.S. Department of Commerce, National Oceanic and Atmospheric Administration, National Marine Fisheries Service, U.S. Department of the Interior Bureau of Land Management, and U.S. Environmental Protection Agency.

Finch, D.M. 1991. Population Ecology, Habitat Requirements and Conservation of Neotropical Migratory Birds. Gen. Tech. Report RM-205. Fort Collins, CO: U.S. Department of Agriculture, Forest Service, Rocky Mountain Forest and Range Experiment Station. 33pp.

Flather, C.H. and T.W. Hoekstra. 1989. An Analysis of the Wildlife and Fish Situation in the United States: 1989-2040. Gen. Tech. Rep. RM-178. Fort Collins, CO: USDA, Forest Service, Rocky Mountain Forest and Range Experimental Station. 147 pp.

Flather, C.H., S.J. Brady and M.S. Knowles. 1999. Wildlife resource trends in the United States: A technical document supporting the 2000 RPA assessment. Gen. Rep. RMRS-GTR-33. Fort Collins, CO: USDA, Forest Service, Rocky Mountain Research Station. 79pp.

Fogel, R. and J.M. Trappe. 1978. Fungus consumption (mycophagy) by small animals. Northwest Sci. 52(1):1-31.

Foley, P. 1997. Extinction models for local populations. Pp. 215-246 in: Metapopulation Biology: Ecology, Genetics, and Evolution. I.A. Hanski and M.E. Gilpin, eds. San Diego, CA: Academic Press.

Force, J.E., G.E. Machlis and L. Zhang. 1994. Understanding Social Change in Resource-Dependent communities. Paper presented at Forestry and the Environment: Economic Perspectives Conference. October 12-15, 1994. Banff, Alberta, Canada.

Force, J.E., G.E. Machlis, L. Zhang, and A. Kearney. 1993. The relationship between timber production, local historical events, and community social change: A quantitative case study. For. Sci. 39(4):722-742.

Forestry Canada. 1993. The State of Canada's Forests, 1992: Third Report to Parliament Forestry Canada, Ottawa. 112 pp.

Forsman, E.D., S. Destefano, M.G. Raphael and R.J. Gutierrez, eds. 1996. Demography of Northern Spotted Owl. Studies in Avian Biology 17. Los Angeles: Cooper Ornithological Society. 122pp.

Frank, S.A. 1997. Spatial processes in host-parasite genetics. Pp. 325-352 in: Metapopulation Biology: Ecology, Genetics, and Evolution, I.A. Hanski and M.E. Gilpin, eds. San Diego: Academic Press.

Frank, D.A. and S.J. McNaughton. 1991. Stability increases with diversity in plant communities: empirical evidence from the 1988 Yellowstone drought. Oikos. 62(3):360-362.

Franklin, J.F. 1979. Vegetation of the Douglas-fir region. Pp. 93-112 in: Forest Soils of the Douglas-Fir Region, P.E. Heilman, H.W. Anderson, and D.M. Baumgartner, eds. Pulman, WA: Washington State University Cooperative Extension Service.

Franklin, J.F. 1991. Old growth and the new forestry. Pp.1-11 in: Proceedings of the New Perspectives Workshop, 1990 July 17-20, Petersburg, Alaska, M.J. Copenhagen, ed. Juneau, AK: USDA, Forest Service, Alaska Region.

Franklin, J.F. 1993b. The fundamentals of ecosystem management with applications in the Pacific Northwest. Pp. 127-144 in: Defining Sustainable Forestry. G.H. Aplet, N. Johnson, J.T. Olson and V.A. Sample, eds. Washington, D.C.: Island Press.

Franklin, J.F. 1993a. Preserving biodiversity: species, ecosystems, and landscapes? Ecological Applications. 3(2):202-205.

Franklin, J.F., and C.T. Dyrness. 1973. Natural Vegetation of Oregon and Washington. Gen. Tech. Rep. PNW-8. Portland, OR: USDA, Forest Service, Pacific Northwest Forest and Range Experiment Station. 417 pp.

Franklin, J.F. and M.A. Hemstrom. 1981. Aspects of succession in the coniferous forest of the Pacific Northwest. Pp. 212-229 in: Forest Succession: Concepts and Application, D.C. West, H.H. Shugart and D.B. Botkin, eds. New York: Springer-Verlag.

Franklin, J.F. and T.A. Spies. 1984. Characteristics of Old-Growth Douglas-fir Forests. Pp. 328-334 in: New Forests For a Changing World. Bethesda, MD: Society of American Foresters.

Franklin, J.F. and T.A. Spies. 1991a. Ecological definitions of old-growth Douglas-fir forests. Pp. 61-71 in: Wildlife and Vegetation of Unmanaged Douglas-Fir Forests. Gen. Tech. Rep. PNW-GTR-285. Portland, OR: USDA, Forest Service, Pacific Northwest Forest and Range Experiment Station.

Franklin, J.F. and T.A. Spies. 1991b. Composition, function, and structure of old-growth Douglas-fir forests. Pp. 71-80 in: Wildlife and Vegetation of Unmanaged Douglas-Fir Forests. Gen. Tech. Rep. PNW-GTR-285. Portland, OR: USDA, Forest Service, Pacific Northwest Forest and Range Experiment Station.

Franklin, J.F., D.R. Berg, D.A. Thornburgh and J.C. Tappeiner. 1997. Alternative silvilcultural approaches to timber harvesting: variable retention harvest systems. Pp. 111-139 in: Creating a Forestry for the 21th Century, K.A. Kohm and J.F. Franklin. Washington, D.C.: Island Press.

Franklin, J.F., K. Cromack, Jr., W. Denison, A. McKee, C. Maser, J. Sedell, F. Swanson and G. Juday. 1981. Ecological Characteristics of Old-Growth Douglas-fir Forests. Gen. Tech. Rep. PNW-118. Portland, OR: USDA, Forest Service, Pacific Northwest Forest and Range Experiment Station. 48pp.

Franklin, J.F., D.A. Perry, T.D. Schowalter, M.E. Harmon, A. McKee and T. Spies. 1989. Importance of ecological diversity in maintaining long-term site productivity. Pp. 82-97 in: Maintaining the Long-Term Productivity of Pacific Northwest Forest Ecosystems, D.A. Perry, R. Meurisse, B. Thomas, R. Miller, J. Boyle, J. Means, C.R. Perry and R.F. Powers, eds. Portland, OR: Timber Press.

Frenkel, R.E. 1993. Vegetation. Pp. 58-65 in: Atlas of the Pacific Northwest, 8th Ed., P.L. Jackson and A.J. Kimerling, eds. Corvallis, Oregon: Oregon State University Press.

Fryer, G.I. and E.A. Johnson. 1988. Reconstructing the fire behaviour and effects in a subalpine forest. J. Appl. Ecol. 25(3):1063-1072.

Fuguitt, G.V. and C.L. Beale. 1993. The changing concentration of the older nonmetropolitan population, 1960-1990. J. Gerontol. 48(6):278-288.

Furniss, R.L. and V.M. Carolin. 1977. Western Forest Insects. USDA For. Serv. Misc. Publ. 1339. Washington, D.C.: GPO.

Galston, W.A. and K.J. Baehler. 1995. Rural Development in the United States: Connecting Theory, Practice and Possibilities. Washington, DC: Island Press. 353pp.

Garrett, M.G., J.W. Watson and R.G. Anthony. 1993. Bald eagle home range and habitat use in the Columbia River estuary. J. Wildl. Manage. 57(1):19-27.

Gates, P.W. 1968. History of Public Land Law Development. Washington, DC.:GPO. 828 pp.

Gibbons, J.W. 1988. The management of amphibians, reptiles and small mammals in North America: the need for an environmental attitude adjustment. Pp. 4-10 in: Management of Amphibians, Reptiles and Small Mammals in North America. Gen. Tech. Rep. RM-166. Fort Collins, CO.: USDA, Forest Service, Rocky Mountain Forest and Range Experiment Station.

Gilpin, M.E. and M.E. Soulè. 1986. Minimum viable populations: Processes of species extinction. Pp. 19-34. In: Conservation Biology: The Science of Scarcity and Diversity, M. Soulè, ed. Sunderland, MA: Sinauer Associates.

Goheen, D.J. and E.M. Hansen. 1993. Effects of pathogens and bark beetles on forests. Pp. 175-196 in: Beetle-Pathogen Interaction in Conifer Forests, T.D. Schowalter and G.M. Filip, eds. London: Academic Press.

Goodman, D. 1975. The theory of diversity-stability relationships in ecology. Q. Rev. Biol. 50(3):237-266.

Gordon, J.C. 1994. From vision to policy: a role for foresters. J. For. 92(7):16-19.

Gower, S.T., C.C. Grier, and K.A. Vogt. 1989. Aboveground production and N and P use by Larix occidentalis and Pinus contorta in the Washington Cascades, USA. Tree Physiol. 5(1):1-11.

Greenstone, M.H. 1984. Determinants of web spider species diversity: vegetation structural diversity vs. prey availability. Oecologia 62(3):299-304.

Grier, C.C. 1975. Wildfire effects on nutrient distribution and leaching in a coniferous ecosystem. Can. J. For. Res. 5(4):599-607.

Grier, C.C. and R.S. Logan. 1977. Old-growth Pseudotsuga menziesii communities of western Oregon watershed: Biomass distribution and production budgets. Ecol. Monogr. 47(4):373-400.

Groves, C. and W. Melquist. 1990. Nongame: Birds, Mammals, Reptiles, Amphibians: Species Management Plan 1991-1995. Boise: Idaho Dept. of Fish and Game. 7pp.

Gruell, G.E. 1985. Indian Fires in the Interior West: A Widespread Influence. Pp. 68-74 in: Gen. Tech. Rep. INT-182. Ogden, UT: USDA, Forest Service, Intermountain Forest and Range Experiment Station.

Grumbine, R.E. 1994. What is ecosystem management? Conserv. Biol. 8(1):27-38.

Hagenstein, P.R. 1992. Some history of multiple use and sustained yield concepts. Pp. 31-43 in: Multiple Use and Sustained Yield: Changing Philosophies for Federal Land Management? Proceedings and Summary of a Workshop. Congressional Research Service, Library of Congress. Committee Print No.11, Committee on Interior and Insular Affairs, U.S. House of Representatives. Washington, D.C.:GPO.

Hagle, S.K. and D.J. Goheen. 1988. Root disease response to stand culture. Pp. 303-309 in: Proceedings-Future Forests of the Mountain West: A Stand Culture Symposium. Gen. Tech. Rep. INT-243. Ogden, UT: USDA, Forest Service, Intermountain Research Station.

Hagle, S. and R. Schmitz. 1993. Managing root disease and bark beetles. Pp. 209-228 in: Beetle-Pathogen Interactions in Conifer Forests, T.D. Schowalter, and G.M. Filip, eds. London: Academic Press.

Hamer, T.E. and S.K. Nelson. 1995. Characteristics of marbled murrelet nest and nesting stands. Pp. 69-82 in: Ecology and Conservation of the Marbled Murrelet, C.J. Ralph, G.L. Hunt, M.G. Raphael and J.F.

Piat, eds. Gen. Tech. Rep. PSW-GTR-152. Albany, California: USDA, Forest Service, Pacific Southwest Research Station.

Hammond, H. 1991. Seeing the Forest Among the Trees: The Case for Wholistic Forest Use. Vancouver, B.C., Canada: Polestar Press.

Hann, W.J., R.E. Keane, C. McNicoll and J. Menakis. 1994. Assessment techniques for evaluating ecosystem processes and community and landscape conditions. Pp. 237-253 in: Eastside Forest Ecosystem Health Assessment, Vol. II: Ecosystem Management: Principles and Applications., M.E. Jensen and P.S. Bourgeron, tech eds. Gen. Tech. Rep. PNW-GTR-318. Portland, OR: USDA, Forest Service, Pacific Northwest Station.

Hansen, A.J., S.L. Garman, J.F. Weigand, D.L. Urban, W.C. McComb, and M.G. Raphael. 1995. Alternative silvicultural regimes in the Pacific Northwest: simulations of ecological and economic effects. Ecological Applications 5(3):535-554.

Hansen, A.J., T.A. Spies, F.J. Swanson, and J.L. Ohmann. 1991. Conserving biodiversity in managed forests. Bioscience. 41(6):382-392.

Hanski, I.A. and M.E. Gilpin, eds. 1997. Metapopulation Biology: Ecology, Genetics and Evolution. San Diego, CA: Academic Press. 512pp.

Hanski, I.A. and D. Simberloff. 1997. The metapopulation approach, its history, conceptual domain, and application to conservation. Pp. 5-26 in: Metapopulation Biology: Ecology, Genetics, and Evolution. I.A. Hanski and M.E. Gilpin, eds. San Diego, CA: Academic Press.

Hanson, F.B. and H.C. Tuckwell. 1981. Logistic growth with random destiny independent disasters. Theor. Popul. Biol. 19(1):1-18.

Harborne, J.B. 1994. Introduction to Ecological Biochemistry, 4th Ed. London: Academic Press.

Harcombe, P.A. 1986. Stand development in a 130-year-old spruce - hemlock forest based on age structure and 50 years of mortality data. For. Ecol. Manage. 14(1):41-58.

Harley, J.L. and S.E. Smith. 1983. Mycorrhizal Symbiosis. London: Academic Press. 483pp.

Harmon, M.E., J.F. Franklin, F.J. Swanson, P. Sollins, S.V. Gregory, J.D. Lattin, N.H. Anderson, S.P. Cline, N.G. Aumen, J.R. Sedell, G.W. Lienkaemper, K. Cromack and K.W. Cummins. 1986. Ecology of coarse woody debris in temperate ecosystems. Adv. Ecol. Res. 15:133-302.

References

Harris, L.D. 1984. The Fragmented Forest: Island Biogeography Theory and the Preservation of Biotic Diversity. Chicago, IL: University of Chicago Press.

Harrison, S. and A.D. Taylor. 1997. Empirical evidence for metapopulation dynamics. Pp. 27-42 in: Metapopulation Biology: Ecology, Genetics, and Evolution, I.A. Hanski and M.E. Gilpin, eds. San Diego, CA: Academic Press.

Harvey, A.E., M.F. Jurgensen and N.J. Larsen. 1978. Role of Residue in and Impacts of its Management of Forest Soil Biology. Proceedings of the 8th World Forestry Congress: Forestry for Quality of Life, Jakarta, 16-18 Oct.1978. FAO Spec. Pap. 11pp.

Haynes, R.W. and D.M. Adams. 1992. The timber situation in the United States-analysis and projections to 2040. J. For. 90(5)38-43.

Haynes, R.W. and J.F. Weigand. 1997. The context for forest economics in the 21st century. In: Creating a Forestry for the 21st Century, K.A. Kohm and J. F. Franklin, eds. Washington D.C.: Island Press.

Haynes, R.W., D.M. Adams and J.R. Mills. 1995. The 1993 RPA Timber Assessment Update. Fort Collins, Co: USDA, Forest Service, Rocky Mountain Forest and Range Experiment Station.

Hayward, G.D., P.H. Haywood and E.O. Garton. 1993. Ecology of Boreal Owls in the Northern Rocky Mountains. Wildl. Monogr.124. Bethesda, MD: Wildlife Society. 59pp.

Heberlein, T.A., R.C. Stedman, G.V. Fuguitt, R.M. Gibson, and P.R. Voss. 1994. Forest Dependence and Community Well Being in the Pacific Northwest. Paper presented at the Annual Meeting of the Rural Sociological Society, August 11-14. Portland, OR. Online. Available: http://www.ssc.wisc.edu/ruralsoc/vosscv.htm

Hedrick, P.W. and M.E. Gilpin. 1997. Genetic effective size of a metapopulation. Pp. 166-181 in: Metapopulation Biology: Ecology, Genetics, and Evolution, I.A. Hanski and M.E. Gilpin, eds. San Diego, CA: Academic Press.

Hemstrom, M.A. and J.F. Franklin. 1982. Fire and other disturbances of the forests in Mount Rainier National Park. Quaternary Research 18(1):32-51.

Henderson, D.M. et al. 1977. Endangered and Threatened Plants of Idaho: A Summary of Current Knowledge. Rare and Endangered Plants Technical Committee, Idaho Natural Areas Council. Bulletin No. 21. Moscow: Forest, Wildlife and Range Experiment Station University of Idaho. 72pp.

Henjum, M.G., J.R. Karr, D.L. Bottom, D.A. Perry, J.C. Bednarz, S.G. Wright, S.A. Beckwitt and E. Beckwitt. 1994. Interim Protection for Late-Successional Forests, Fisheries and Watersheds: National Forests East of the Cascade Crest, Oregon and Washington, J.R. Karr and E.W. Chu, eds. Bethesda, MD: The Wildlife Society. 235pp.

Hepting, G.H. 1971. Diseases of Forest and Shade Trees of the United States. USDA Forest Service Agriculture Handbook No. 386. Washington, D.C.: USDA, Forest Service. 658pp.

Hessburg, P., M. Jensen, B. Borman and R. Everett. 1993. Eastside Forest Ecosystem Health Assessment. National Forest System, Forest Service Research. USDA. April,1993.

Hessburg, P.F., R.G. Mitchell and G.M. Filip. 1994. Historical and Current Roles of Insects and Pathogens in Eastern Oregon and Washington Forested Landscapes. Gen. Tech. Rep. PNW 317. Portland, OR: USDA, Forest Service, Pacific Northwest Research Station.

Hibbard, M. and J. Elias. 1993. The failure of sustained-yield forestry and the decline of the flannel-shirt frontier. Pp. 195-217 in: Forgotten Places: Uneven Development in Rural America, T.A. Lyson, and W.W. Falk, eds. Lawrence: University of Kansas Press.

Hicks, J.R. 1946. Value and Capital: An Inquiry into Some Fundamental Principles of Economic Theory, 2nd Ed. London: Oxford University Press. 340 pp.

Hilborn, R. 1992. Hatcheries and the future of salmon in the northwest. Fisheries 17(1):5-8.

Hilborn, R. and M. Mangel. 1997. The Ecological Detective: Confronting Models With Data. Princeton, N.J.: Princeton University Press.

Hof, J. and M.G. Raphael. 1997. Optimization of habitat placement: a case study of the Northern spotted owl in the Olympic Peninsula. Ecological Applications 7(4):1160-1169.

Holt, D.W. and J.M. Hillis. 1987. Current status and habitat associations of forest owls in western Montana. Pp. 281-288 in: Biology and Conservation of Northern Forest Owls. R.W. Nero et al, eds. Gen. Tech. Rep. RM-142. Fort Collins, Co: USDA, Forest Service, Rocky Mountain Forest and Range Experiment Station.

Howard, J.L. 1999. U.S. Timber Production, Trade, Consumption, and Price Statistics 1965-1997. Gen. Tech. Report FPL-GTR-116. Madison WI: USDA. Forest Service, Forest Products Laboratory.

Howze, G., C. Bailey, J. Bliss, and L. Teeter. 1993. Regional Comparisons of Timber Dependency: The Northwest and the Southeast. Paper presented at the Annual Meeting of the Rural Sociological Society, August 7-10, 1993, Orlando, FL.

Huff, M.H. 1984. Post-Fire Succession in the Olympic Mountains, Washington: Forest Vegetation, Fuels and Avifauna. Ph.D. dissertation. University of Washington, Seattle.

Humphrey et al. 1993. Theories in the study of natural resource dependent communities and persistent rural poverty in the United States. Pp. 136-172 in: Persistent Poverty in Rural America. Rural Sociological Society Task Force on Persistent Rural Poverty. Boulder: Westview.

Hunter, A.F. and L.W. Aarssen. 1988. Plants helping plants. Bioscience 38(1):34-40.

Hutchison, S.B. And R. K. Winters. 1942. Northern Idaho Forest Resources and Industries. Washington, D.C.: U.S. Govt. Print. Off. 75pp.

IFMAT (Indian Forest Management Assessment Team). 1993. An Assessment of Indian Forest and Forest Management in the United States. Portland, OR: the Intertribal Timber Council.

Ince, P.J. 1994. Recycling and Long-Range Timber Outlook. Gen. Tech. Rep. RM-242. Fort. Collins, CO: USDA, Forest Service, Rocky Mountain Forest and Range Experiment Station. 23 pp.

Irwin, L.L., and T.L. Fleming, eds. 1991. Demography of Spotted Owls in Washington's Eastern Cascades, 1990 Annual Progress Report. Corvallis, OR: National Council of the Paper Industry for Air and Stream Improvement, Inc.

Jackson, D.H. and K.O. Jackson. 1987. An economic analysis of production and markets: the post and pole sector in Montana. Pp. 83-84 in: Gen. Tech. Rep. INT-237. USDA, Forest Service, Intermountain Research Station.

Johnson, K.H., K.A. Vogt, H.J. Clark, O.J. Schmitz, and D.J. Vogt. 1996. Biodiversity and the productivity and stability of ecosystems. Tree. 11:372-377.

Johnson, K.N., J.F. Franklin, J.W. Thomas, and J. Gordon. 1991. Alternatives for Management of Late-Successional Forests of the Pacific Northwest. A report to the Agriculture Committee and the Merchant Marine Committee of the U.S. House of Representatives.

Kareiva, P. 1994. Diversity begets productivity. Nature 368(6473):686-687.

Kareiva, P. 1983. Influence of vegetation texture on herbivore populations: resource concentration and herbivore movement. Pp. 259-289 in: Variable Plants and Herbivores in Natural and Managed Ecosystems, R.F. Denno and M.S. McClure, eds. New York: Academic Press.

Kessler, W.B., H. Salwasser, and C.W. Cartwright, Jr. 1992. New perspective for sustainable natural resource management. Ecological Applications 2(3):221-225.

Klein, B.C. 1989. Effects of forest fragmentation on dung and carrion beetle communities in central Amazonia. Ecology 70(6):1715-1725.

Koch, P. 1992. Wood versus nonwood materials in US residential construction: some energy-related global implications. Forest Products Journal 42(5):31-42.

Kohm, K.A. and J.F. Franklin, eds. 1997. Creating a Forestry for the 21st Century. Washington, D.C.:Island Press.

Konkel, G.W. and J.D. McIntyre. 1987. Trends in Spawning Populations of Pacific Anadromous Salmonids. Fish and Wildlife Tech. Rep. 9, U.S. Dept. of Interior, Fish and Wildlife Service, Washington, D.C.

Koricheva, J., S. Larsson and E. Haukioja. 1998. Insect performance on experimentally stressed woody plants: a meta-analysis. Annu. Rev. Entomol. 43:195-216.

Kruess, A. and T. Tscharntke. 1994. Habitat fragmentation, species loss and biological control. Science 264(5165):1581-1584.

Kuck, L., ed. 1992. Idaho Department of Fish and Game Statewide Surveys and Inventory: Elk. Project W170R16, Study I, Job 1. Boise, Idaho.

Kuck, L., ed. 1993. Idaho Department of Fish and Game Statewide Surveys and Inventory: Elk. Project W170R17. Study I, Job 1. Boise, Idaho.

Kusel, J. and L. Fortmann. 1991. Well-Being in Forest-Dependent Communities. 1991. Sacramento, CA: California Department of Forestry and Fire Protection, Forest and Rangeland Resources Assessment Program (FRRAP).

Lande, R. 1988. Demographic models of the northern spotted owl. (Strix occidentalis caurina). Oecologia 75(4):601-607.

Lande, R. 1993. Risks of population extinction from demographic and environmental stochasticity and random catastrophes. Am. Nat. 142(6):911-927.

Larsen, J.A. 1930. Forest types of the Northern Rocky Mountains and their climatic controls. Ecology 11:631-672.

Lattin, J.D. 1990. Arthropod diversity in Northwest old-growth forests. Wings 15(2):7-10.

Ledig, F.T. 1986. Conservation strategies for forest gene resources. For. Ecol. Manage. 14(2):77-90.

Lee, K.J. and F.W. Cubbage. 1993. Timber Dependency in Georgia. Paper presented at the Annual Meeting of the Rural Sociological Society, August 7-10, 1993. Orlando, FL.

Lee, R.G. 1993. A Constructive Critique of the FEMAT Social Assessment. An Independent Paper prepared for The American Forest and Paper Association, California Forestry Association, Northwest Forestry Association. October, 15.

Lehmkuhl, J.F. and M.G. Raphael. 1993. Habitat pattern around spotted owl location on the Olympic Peninsula, Washington. J. Wildl. Manage. 57(2):302-315.

Lehmkuhl, J.F., P.F. Hessburg and R.L. Everett. 1994. Historical and Current Forest Landscapes of Eastern Oregon and Washington. Part I: Vegetation Pattern and Insect and Disease Hazards. Gen. Tech. Rep. PNW-GTR-328. Portland, OR: USDA, Forest Service, Pacific Northwest Research Station. 88pp.

Leps, J., J. Osbornova-Kosinova and M. Rejmanek. 1982. Community stability, complexity and species life history strategies. Vegetatio. 50(1):53-63.

Leshy, J.D. 1992. Is the multiple use, sustained yield management philosophy still applicable today? Pp.107-119 in: Multiple Use and Sustained Yield: Changing Philosophies for Federal Land Management? Proceedings and Summary of a Workshop. Congressional Research Service, Library of Congress. Committee Print No.11, Committee on Interior and Insular Affairs, U.S. House of Representatives. Washington D.C.:GPO.

Levin, S.A. 1992. MacArthur Award Lecture: The problem of pattern and scale in ecology. Ecology 73(6):1943-1967.

Lloyd, J.D., Jr., J. Moen, and C.L. Bolsinger. 1986. Timber resource statistics for the north coast resource area of California. Resource Bulletin PNW 131. USDA Forest Service Pacific Northwest Forest and Range Experiment Station.

Lomolino, M.V., J.H. Brown and R. Davis. 1989. Island biogeography of Montane forest mammals in the American Southwest. Ecology 70(1):180-194.

Lorio, P.L., Jr. 1993. Environmental stress and whole-tree physiology.

Pp. 81-101 in: Beetle-Pathogen Interaction in Conifer Forests, T.D. Schowalter and G.M. Filip, eds. London: Academic Press.

Lyon, L.J. 1984. The Sleeping Child Burn-21 Years of Post-Fire Change. Res. Pap. INT 330. USDA, Forest Service, Intermountain Forest and Range Experiment Station. 17pp.

MacArthur, R.H., and E.O. Wilson. 1967. The Theory of Island Biogeography. Princeton, N.J.: Princeton University Press. 203pp.

MacCleery, D. 1995. American Forests: A History of Resiliency and Recovery. Forest History Society Monograph, Durham, N.C.

Malmquist, M.G. 1985. Character displacement and biogeography of the pygmy shrew in Northern Europe. Ecology 66(2):372-377.

Manion, P.D. 1981. Tree Disease Concepts. Englewood Cliffs, NJ: Prentice-Hall.

Marquis, R.J. and C.J. Whelan. 1994. Insectivorous birds increase growth of white oak through consumption of leaf-chewing insects. Ecology 75(7):2007-2014.

Martin, R.E. 1982. Fire history and its role in succession. Pp. 92-99 in: Forest Succession and Stand Development Research in the Northwest, J.E. Means, ed. Corvallis, OR.: Forest Research Laboratory, Oregon State Univ.

Maser, C. 1990. The Redesigned Forest. Toronto, Canada: Stoddart.

Maser, C., J.M. Trappe and R.A. Nussbaum. 1978. Fungal-small mammal interrelationships with emphasis on Oregon coniferous forests. Ecology 59(4):799-809.

Maser, C., R.G. Anderson, K. Cromack, Jr., J.T. Williams and R.E. Martin. 1979. Dead and down woody material. Pp. 78-95 in: Wildlife Habitats in Managed Forests, the Blue Mountains of Oregon and Washington., J.Thomas, J.L. Parker, R.A. Mowrey, G.M. Hanson and B.J. Bell. eds. Agriculture Handbook 553. Washington, D.C: USDA, Forest Service.

Mattson, W.J. 1980. Herbivory in relation to plant nitrogen content. Annu. Rev. Ecol. Syst. 11:119-161.

Mattson, W.J. and R.A. Haack. 1987. The rolle of drought in outbreaks of plant-eating insects. Bioscience. 37:110-118.

May, R.M. 1973. Stability and Complexity in Model Ecosystems. Princeton University Press. 235pp.

McClelland, B.R. 1979. The pileated woodpecker inforests of the northern Rocky Mountains. Pp. 283-299 in: The Role of Insectivorous

Birds in Forest ecosystems, J.G. Dickson, R.N. Conner, R.R. Fleet, J.C. Kroll and J.A. Jackson, eds. New York: Academic Press.

McClelland, B.R., S.S. Frissell, W.C. Fischer and C.H. Halvorson. 1979. Habitat management for hole-nesting birds in forests of western larch and Douglas-fir. J. For. 77(8):480-483.

McComb, W., T.A. Spies and W.H. Emmingham. 1993. Douglas-fir forests: managing for timber and mature-forest habitat. J. For. 91(12):31-42.

McComb, W., J. Tappeiner, L. Kellogg, R. Johnson, and C. Chambers. 1994. Stand management alternatives for multiple resources: integrated management experiments. Pp. 71-86 in: Expanding Horizons of Forest Ecosystem Management: Proceedings of the Third Habitats Futures Workshop, M.H. Huff, S. E. McDonald, and H. Gucinski, eds. Gen. Tech. Rep. PNW-GTR-336. Portland, OR: USDA, Forest Service, Pacific Northwest Research Station.

McCool, S.F., A.E. Watson, et al. 1995. Linking Tourism, the Environment, and Sustainability . Gen. Tech. Report. INT-323. Ogden, UT: USDA, Forest Service, Intermountain Research Station.

McGarigal, K. and W. C. McComb. 1995. Relationships between landscape structure and breeding birds in the Oregon Coast Range. Ecol. Monogr. 65(3):235-260.

McNaughton, S.J. 1977. Diversity and stability of ecological communities: a comment on the role of empiricism in ecology. Am. Nat. 111(979):515-525.

McNaughton, S.J. 1985. Ecology of a grazing ecosystem: The Serengeti. Ecol. Monogr. 55(3):259-294.

Merrill, E.H., H.F. Mayland and J.M. Peek. 1980. Effects of a fall wildfire on herbaceous vegetation on xeric sites in the Selway-Bitterroot wilderness, Idaho. J. Range Manage. 33(5):363-367.

Meslow, E.C. and H.M. Wight. 1975. Avifauna and succession in Douglas-fir forests of the Pacific Northwest. Pp. 266-271 in: Proceedings of the Symposium on Management of Forest and Range Habitats for Nongame Birds. Gen. Tech. Rep. WO-1. Washington, D.C.: USDA, Forest Service.

Meyer, J.S., L.L. Irwin, and M. S. Boyce. 1998. Influence of habitat abundance and fragmentation on nothern spotted owls in western Oregon. Wildlife Monographs No 139. July 1998. 51 pp.

Mills, L.S., M.E. Soulé, and D.F. Doak. 1993. The keystone-species concept in ecology and conservation. Bioscience 43(4):221-224.

Milne, B.T. and R.T. Forman. 1986. Peninsulas in Maine: woody plant diversity, distance and environmental patterns. Ecology 67(4):967-974.

Molina, R., and J.M. Trappe. 1982. Patterns of ectomycorrhizal host specificity and potential among Pacific Northwest conifers and fungi. For. Sci. 28(3):423-458.

Molina, R., T. O'Dell, D. Luoma, M. Amaranthus, M. Castellano, K. Russell. 1993. Biology, Ecology, and Social Aspects of Wild Edible Mushrooms in the Forests of the Pacific Northwest: A Preface to Managing Commercial Harvests. Gen. Tech. Rep. PNW-GTR-309. Portland, OR: USDA, Forest Service, Pacific Northwest Research. 42 pp.

Molina, R., N. Vance, J. F. Weigand, D. Pilz, and M. P. Amaranthus. 1997. Special forest products: integrating social, economic, and biological considerations into ecosystem management. Pp. 315-336 in: Creating a Forestry for the 21st Century, K.A. Kohm and J. F. Franklin, eds. Washington DC.: Island Press.

Molles, M.C. 1978. Fish species diversity on model and natural reef patches: experimental insular biogeography. Ecol. Monogr. 48:289-305.

Montana Bald Eagle Working Group. 1991. Habitat Management Guide for Bald Eagles in Northwestern Montana. Billings, MT: USDI Bureau of Land Management. 29pp.

Morgan, P. and L.F. Neuenschwander. 1988. Shrub response to high and low severity burns following clearcutting in northern Idaho. Western J. Appl. For. 3(1):5-9.

Morrison, P.H. 1991. Ancient forests in the Pacific Northwest, analysis and maps of twelve national forests in: Ancient Forests Existing in 1989 and Northern Spotted Owl Habitat Conservation Areas. Washington, D.C.: Wilderness Society.

Morrison, P.H. and F.J. Swanson. 1990. Fire History and Pattern in a Cascade Range Landscape. Gen. Tech. Rep. PNW-GTR-254. Portland, OR: USDA, Forest Service, Pacific Northwest Research Station.

Moulton, R.J. 1998. Tree planting in the United States – 1997. Tree Planters' Notes 49:1-15.

Moulton, R.J., R.D. Mangold, and J.D. Snellgrove. 1993. Tree Planting

in the United States, 1992. Washington, D.C.: USDA, Forest Service. 15 pp.
Mueggler, W.F. 1965. Ecology of seral shrub communities in the cedar-hemlock zone of northern Idaho. Ecol. Monogr. 25:165-185.
Mutch, R.W., S.F. Arno, J.K. Brown, C.E. Carlson, R.D. Ottmar and J.L. Peterson. 1993. Forest Health in the Blue Mountains. A Management Strategy for Fire-Adapted Ecosystems. Gen. Tech. Rep. PNW-GTR-310. Portland, OR: USDA, Forest Service, Pacific Northwest Research. 14pp.
Naeem, S., L.J. Thompson, S.P. Lawler, J.H. Lawton and R.M. Woodfin. 1994. Declining biodiversity can alter the performance or ecosystems. Nature 368(6473):734-737.
Nash, R. 1982. Wilderness and the American Mind, 3rd Ed. New Haven, Conn: Yale University Press. 425pp.
National Association of Counties. 1993. America's Endangered Communities. Washington D.C.: National Association of Counties.
Neitlich, P.N. and B. McCune. 1997. Hotspots of epiphytic lichen diversity in two young managed forests. Conserv. Biol. 11(1):172-182.
Nel, E.M., C.A. Wessman and T.T. Veblen. 1994. Digital and visual analysis of thematic mapper imagery for differentiating old growth from younger spruce-fir stands. Remote Sens. Environ. 48(3):291-301.
Nelson, S.K., M.L.C. McAllister, M.A. Stern, D.H. Varoujean and J.M. Scott. 1992. The marbled murrelet in Oregon, 1899-1987. Proc. Western Foundation of Vertebrate Zoology. 5(1):61-91.
Neuberger, R.L. 1938. Our Promised Land. New York: MacMillan. 398 pp.
Nilsson, S.G. and I.N. Nilsson. 1978. Species richness and dispersal of vascular plants to islands in Lake Mockeln, Southern Sweden. Ecology 59(3):473-480.
Noon, B.R. and C.M. Biles. 1990. Mathematical demography of spotted owls in the Pacific Northwest. J. Wildl. Manage. 54(1):18-27.
Norse, E.A. 1990. Ancient Forests of the Pacific Northwest. Washington, D.C.: Island Press. 327pp.
Northwest Power Planning Council. 1986. Compilation of Information on Salmon and Steelhead Losses in the Columbia River Basin. Portland, OR: Northwest Power Planning Council.

Noss, R.F. 1983. A regional landscape approach to maintain diversity. BioScience 33(11):700-706.

Noss, R.F. and A.Y. Cooperrider. 1994. Saving Nature's Legacy: Protecting and Restoring Biodiversity. Washington, D.C.: Island Press. 416pp.

NRC (National Research Council). 1976. Renewable Resources for Industrial Materials. Washington DC: National Academy of Sciences. 266 pp.

NRC (National Research Council). 1990. Forestry Research. A Mandate for Change. National Academy Press: Washington D.C. 84pp.

NRC (National Research Council). 1993. Setting Priorities for Land Conservation. National Academy Press: Washington, D.C. 262pp.

NRC (National Research Council). 1994. Rangeland Health: New Methods to Classify, Inventory, and Monitor Rangelands. Washington, D.C: National Academy Press. 200pp.

NRC (National Research Council). 1995. Science and the Endangered Species Act. Washington, D.C.: National Academy Press. 288pp.

NRC (National Research Council). 1996. Upstream: Salmon and Society in the Pacific Northwest. Washington, D.C: National Academy Press. 452pp.

NRC (National Research Council). 1998. Forested Landscapes in Perspective: Prospects and Opportunities for Sustainable Management of America's Nonfederal Forests. Washington, D.C: National Academy Press. 249pp.

NRC (National Research Council). 1999a. Our Common Journey: A Transition Toward Sustainability. Washington D.C.: National Academy Press. 384pp.

NRC (National Research Council). 1999b. Perspectives on Biodiversity, Valuing Its Role in an Everchanging World. Washington, D.C.: National Academy Press. 168pp.

Odum, E.P. 1969. The strategy of ecosystem development. Science 164(3877):262-270.

O'Laughlin, J. 1994. Assessing forest health conditions in Idaho with forest inventory data. Journal of Sustainable Forestry. 2(3/4):221-247.

O'Laughlin, J., J.G. MacCracken, D.L. Adams, S.C. Bunting, K.A. Blatner, and C.E. Keegan, III. 1993. Forest Health Conditions in Idaho. Idaho Forest, Wildlife and Range Policy Analysis Group Report No. 11, University of Idaho. 244 pp.

References

Old-Growth Definition Task Group. 1986. Interim Definition for Old-Growth Douglas-Fir and Mixed-Conifer Forests in the Pacific Northwest and California. Research Note PNW-447. Portland, OR: USDA, Forest Service, Pacific Northwest Research Station. 7pp.

Oliver, C.D. 1981. Forest development in North America following major disturbances. For. Ecol. Manage. 3(3):153-168.

Oliver, C.D. and B.C. Larson. 1996. Forest Stand Dynamics. New York: John Wiley and Sons.

Oliver, C., D. Adams, T. Bonnicksen, J. Bowyer, F. Cubbage, N. Sampson, S. Schlarbaum, R. Whaley and H. Wiant. 1997. Report on Forest Health of the United States by the Forest Health Science Panel. Seattle, Wa: Center for international Trade in forest Products, University of Washington.

Oliver, C.D., D.E. Ferguson, A.E. Harvey, H.S. Malany, J.M. Mandzak and R.W. Mutch. 1994. Managing ecosystems for health: An approach and the effects on uses and values. J. Sustainable For. 2(1/2):133-133.

Oregon Department of Forestry. 1999. First Approximation Report for Sustainable Forest Management in Oregon. Draft. Oregon Department of Forestry. September. Available: http://www.odf.state.or.us/FAR/first%approximation%20report.htm

OTA (Office of Technology Assessment). 1993a. Harmful Non-Indigenous Species in the United States. U.S. Congress, Office of Technology Assessment, OTA-F-565. Washington D.C.: GPO.

OTA (Office of Technology Assessment). 1993b. Wood Use: U.S. Competitiveness and Technology. U.S. Congress, Office of Technology Assessment. Washington, D.C.: GPO.

Overdevest, C. and G.P. Green. 1994. Forest dependency and community well-being: A segmented market approach. Society and Natural Resources. 8(2):111-131.

PACFISH (U.S. Department of Agriculture, U.S. Department of Interior, Bureau of Land Management.). 1994. Environmental Assessment for the Implementation of Interim Strategies for Managing Anadromous Fish-Producing Watersheds in Eastern Oregon and Washington, Idaho, and Portions of California. Washington, D.C.: U.S. Department of Agriculture, U.S. Department of Interior, Bureau of Land Management.

Pacific Seabird Group. 1993. White paper on the status of marbled murrelets. (Unpubl. rep. on file). Pacific Seabird Group, Seattle, WA. 8pp.

Paine, T.D., R..A. Redak and J.T. Trumble. 1993. Impact of acidic deposition on Encelia farinosa Gray (Compositae: Asteraceae) and feeding preferences of Trirhabda geminata Horn (Coleoptera: Chrysomelidae). J. Chem. Ecol. 19(1):97-105.

Palazzi, L.M., R.F. Powers and D.H. McNabb. 1992. Geology and soils. Pp. 48-72 in: Reforestation Practices in Southwestern Oregon and Northern California, S.D. Hobbs, S.D. Tesch, P.W. Owston, R.E. Stewart, J.C. Tappeiner II and G.E. Wells, eds. Corvallis, OR: Forest Research Laboratory.

Parke, J.L., R.G. Linderman, and J.M. Trappe. 1983. Effects of forest litter on mycorrhiza development and growth of Douglas-fir and western red cedar seedlings. Can. J. For. Res. 13(4):666-671.

Parsons, G.L., G. Cassis, A.R. Moldenke, J.D. Lattin, N.H. Anderson, J.C. Miller, P. Hammond, and T.D. Schowalter. 1991. Invertebrates of the H.J. Andrews Experimental Forest, Western Cascade Range, Oregon: V. An annotated list of insects and other arthropods. Gen. Tech. Rep. PNW-GTR-290. Portland, OR: USDA, Forest Service, Pacific Northwest Forest and Range Experiment Station.

Patterson, S. 1992. Douglas-Fir Beetle: Dealing with an Epidemic. Pp. 73-76 in: Gen. Tech. Report INT 291. USDA, Forest Service, Intermountain Research Station.

Paulson, D.R. 1992. Northwest bird diversity: from extravagant past and changing present to precarious future. Northwest Environ. J. 8(1):71-118.

Peck, J.E. and B. McCune. 1997. Remnant trees and conopy lichen communities in western Oregon: a retrospective approach. Ecol. Applic. 7(4):1181-1187.

Peek, J.M.,D.J. Pierce, D.C. Graham, and D.L. Davis. 1987. Moose habitat use and implications for forest management in north central Idaho. Swedish Wildl. Res. (Suppl.1):195-199.

Peet, R.K. and N.L. Christansen. 1987. Competition and tree death. BioScience 37(8):586-595.

Perez-Garcia, J.M. 1993. Global Forestry Impacts of Reducing Softwood Supplies from North America, CINTRAFOR Working Paper 43, May 14, 1993. Draft. University of Washington, Seattle, WA. 39 pp.

Perry, D.A. 1988a. Landscape pattern and forest pests. Northwest Environ. J. 4(2):213-228.
Perry, D.A. 1988b. An overview of sustainable forestry. Journal of Pesticide Reform. 8(3):8-12.
Perry, D.A. 1994. Forest Ecosystems. Baltimore: John Hopkins University Press. 649pp.
Perry, D.A. 1995a. Landscapes, humans, and other system-level considerations: a discourse on ecstasy and laundry. Pp. 177-191 in: Ecosystem Management in Western Interior Forests Symposium Proceedings, May 3-5, 1994, Spokane, WA. Pulman, WA: Washington State University.
Perry, D.A. 1995b. Self-organizing systems across scales. Trends Ecol. Evol. 10(6):241-244.
Perry, D. A. 1998. The scientific basis of forestry. Annu. Rev. Ecol. Syst. 29:435-466.
Perry, D.A. and J. Maghembe. 1989. Ecosystem concepts and current trends in forest management: time for reappraisal. For. Ecol. Manage. 26(2):123-140.
Perry, D.A. and G.B. Pitman. 1983. Genetic and environmental influences in host resistance to herbivory: Douglas-fir and the western spruce budworm. Zeitschrift fur angevandte Entomologie 96(3):217-228.
Perry, D. A., T. Bell and M.P. Amaranthus. 1992. Mycorrhizal Fungi in Miixed-Species Forests and Other Tales of Positive Feedback, Redundancy and Stability. Special pub...of British Ecological Society. 11:151-179. In: the series analytic: The Ecology of Miixed-Species Stands of Trees, M.G.R. Cannell, D.C. Malcolm and P.A. Robertson.
Perry, D.A., R. Meurisse, B. Thomas, R. Miller, J. Boyle, J. Means, C.R. Perry and R.F. Powers, eds. 1989a. Maintaining the Long-Term Productivity of Pacific Forest Ecosystem. Portland, OR: Timber Press.
Perry, D.A., M.P. Amaranthus, J.G. Borchers, S.L. Borchers and R.E. Brainard. 1989b. Bootstrapping in ecosystems. BioScience 39(4):230-237.
Pianka, E.R. 1967. On lizard species diversity: North American flatland deserts. Ecology 48(3):333-351.
Pierce, D.J., B.W. Ritchie and L. Kuck. 1985. An examination of unregulated harvest of Shiras moose in Idaho. Alces 21:231-252.
Pilz, D. and R. Molina. 1996. Managing Forest Ecosystems to Conserve

Fungus Diversity and Sustain Wild Mushroom Harvests. Gen. Tech. Rep. PNW - GTR-371. Portland, OR: USDA, Forest Service, Pacific Northwest Research Station.

Pinchot, G. 1907. The Use of the National Forests. (Reprinted SAF). 42pp.

Potter, D.R., J.C. Hendee, and R.H. Clark. 1973. Hunting satisfaction: game, guns, or nature? Transactions North American Wildlife and Natural Resources Conference 38:220-229.

Powell, A.H. and G.V.N. Powell. 1987. Population dynamics of male euglossine bees in Amazonian forest fragments. Biotropica 19(2):176-179.

Powel, J.H., Jr., and G.K. Loth. 1981. An Economic Analysis of Nontimber Uses of Forest Land in the Pacific Northwest. Final Report. Forest Policy Project. NTIS PB82 109182. Renton, WA: Washington State University.

Powell, D.S., J.L. Faulkner, D.R. Darr, Z. Zhu, and D.W. MacCleery. 1993. Forest Resources of the United States, 1992. Gen. Techn. Rep. RM-234. Fort Collins CO: USDA, Forest Service. 132 pp.

Power, D.M. 1975. Similarity among avifaunas of the Galapagos Islands. Ecology 56(3):616-626.

Power, T.M. 1992. The Economics of Wildland Preservation: The View from the Local Economy. Gen. Tech. Rep. SE 78. USDA, Forest Service. Southeastern Forest Experiment Station.

Power, T.M. 1996. Lost Landscapes and Failed Economies: The Search for a Value of Place. Washington, D.C.: Island Press. 304pp.

Price, M.L. and D.C. Clay. 1980. Structural disturbances in rural communities: some repercussions of the migration turnaround in Michigan. Rural Sociol. 45(4):591-607.

Progar, R.A., T.D. Schowalter, C.M. Freitag and J.J. Morrell. in press. Respiration from coarse woody debris as affected by moisture and saprotroph functional diversity in western Oregon. Oecologia

Punttila, P. Y. Haila, N. Niemela and T. Pajunen. 1994. Ant communities in fragments of old-growth taiga and managed surroundings. Annales Zoologici Fennici 31(1):131-144.

Pyne, S.J. 1982. Fire in America: A Cultural History of Wildland and Rural Fire. Princeton, N.J.: Princeton University Press. 654pp.

Pyne, S.J. 1995. World Fire: The Culture of Fire on Earth, First Ed. New York: Holt.

Radtke, H.D. and S.W. Davis. 1988. The Economic Landscape of the Oregon Coast. Newport OR: Oregon Coastal Zone Management Association. 23 pp.

Ralph, C.J. G.L. Hunt, M.G. Raphael and J.F. Piat, eds. 1995. Ecology and Conservation of the Marbled Murrelet. Gen. Tech. Rep. PSW-GTR-152. Albany, California: USDA, Forest Service, Pacific Southwest Research Station.

Raphael, M.G. 1988. Long-term trends in abundance of amphibians, reptiles and mammals in Douglas-fir forests of northwestern California. Pp. 23-30 in: Management of Amphibians, Reptiles and Small Mammals in North America. Gen. Tech. Rep. RM-166. Fort Collins, CO.: USDA, Forest Service, Rocky Mountain Forest and Range Experiment Station.

Rapport, D.J. 1989. What constitutes ecosystem health? Perspect. Biol. Med. 33(1):120-132.

Rebertus, A.J., T.T. Veblen, L.M. Roovers, and J.N. Mast. 1992. Structure and dynamics of old-growth Engelmann spruce- subalpine fire in Colorado. Pp. 51-59 in: Old-Growth Forest in the Southwest and Rocky Mountain Regions: 1992 March 9-13, Portal, Arizona. M.R. Kaufmann, W.H. Moir and W.R. Bassett, tech. coords. Gen. Tech Rep. RM-213. Fort Collins, CO: USDA, Forest Service, Rocky Mountain Forest and Range Experiment Station.

Reiter, M.L. and R.L. Beschta. 1995. Effects of forest practices on water. Chapter 7 in: Cumulative Effects of Forest Practices in Oregon: Literature and Synthesis. Report for Oregon Department of Forestry. R.L. Beschta, J.R. Boyle, C.C. Chambers, W.P. Gibson, S.V. Gregory, J. Grizzel, J.C. Hager, J.L. Li, W.C. McComb, T.W. Parzybok, M.L. Reiter, G.H. Taylor and J.E. Warila, eds. Corvallis, Oregon: Oregon State University.

Rey, J.R. 1981. Ecological biogeography of arthropods on Spartina Islands in northwest Florida. Ecol. Monogr. 51(2):237-265.

Rhoades, D.F. 1983. Herbivore population dynamics and plant chemistry. Pp. 155-220 in: Variable Plants and Herbivores in Natural and Managed Ecosystems, R.F. Denno and M.S. McClure, eds. New York: Academic Press.

Ribe, R.G. 1989. The aesthetics of forestry: What has empirical preference research taught us? Environ. Manage. 13(1):55-74.

Richardson, E.R. 1980. BLM's Billion-Dollar Checkerboard: Managing the O and C Lands. Santa Cruz, CA.: Forest History Society. 200 pp.

Ripple, W.J. 1994. Historic spatial patterns of old forests in Western Oregon. J. For. 92(11):45-49.

Risch, S. 1980. The population dynamics of several herbivorous beetles in a tropical agroecosystem: the effect of intercropping corn, beans and squash in Costa Rica. J. Appl. Ecol. 17(3):593-611.

Risch, S.J. 1981. Insect herbivore abundance in tropical monocultures and polycultures: an experimental test of two hypotheses. Ecology 62(5):1325-1340.

Robbins, C.S., B. Bruun and H.S. Zim. 1983. Birds of North America: A Guide to Field Identification. New York: Golden Press. 360pp.

Robbins, W.G. 1985. American Forestry: A History of National, State, and Private Cooperation. Lincoln: University of Nebraska Press.

Rochelle, J.A., L.A. Lehmann and J. Wisniewski Rochelle, eds. 1999. Forest Fragmentation: Wildlife and Management Implications. Leiden, Netherlands: Brill.

Rosenzweig, M.L. 1975. On continental steady states of species diversity. Pp.121-140 in: Ecology and Evolution of Communities, M.L. Cody and J.M. Diamond, eds. Cambridge, Massachusetts: The Belknap Press of Harvard University Press.

RSS (Rural Sociological Society Task Force on Persistent Rural Poverty). 1993. Persistent Poverty in Rural America. Boulder: Westview Press.

Rudzitis, G. 1993. Nonmetropolitan geography: migration, sense of place, and the American West. Urban Geography 14(6):574-585.

Rudzitis, G. and H.E. Johansen. 1991. How important is wilderness? Results from a United States survey. Environ. Manage. 15(2):227-233.

Ruggiero, L.F., K.B. Aubry, S.W. Buskirk, L.S. Lyon and W.J. Zielinski. 1994. The Scientific Basis for Conserving Forest Carnivores: American Marten, Fisher, Lynx and Wolverine in the Western United States. Gen. Tech. Rep. RM 254. Fort Collins, CO: USDA, Forest Service, Rocky Mountain Forest and Range Experiment Station.

Ruth, R.H. and A.S. Harris. 1979. Management of Western Hemlosk-Sitka Spruce Forests for Timber Production. Gen. Tech Rep. PNW-88, Portland. OR.: USDA, Forest Service, Pacific Northwest Forest and Range Experiment Station.

Rydin, H. and S. Borgegard. 1988. Plant species richness on islands over a century of primary succession: Lake Hjalmaren. Ecology 69(4):916-927.

SAF (Society of American Foresters). 1993. Sustaining Long-Term

Forest Health and Productivity. Bethesda, MD: Society of American Foresters.

Salazar, D.J., C.H. Schallau and R.G. Lee. 1986. The Growing Importance of Retirement Income in Timber-Dependent Areas. Research paper PNW 359. Portland, OR: USDA, Forest Service, Pacific Northwest Forest and Range Experiment Station.

Salwasser, H. 1990. Gaining perspective: forestry in the future. J. For. 88(11):32-38.

Salwasser, H. 1991. Some perspectives on people, wood, and ecological thinking in forest conservation: Why all the fuss about forest? Pp. 12-20 in: Southwestern Mosaic: Proceedings of the Southwestern Region New Perspective University Coloquium, D.C. Hayes, J.S. Bumstead and M.T. Richards, eds. Gen. Tech. Rep. RM 216. Fort Collins, CO: USDA, Forest Service, Rocky Mountain Forest and Range Experiment Station.

Sample, V.A. and D.C. Le Master. 1992. Economic effect of northern spotted owl protection. J. For. 90(8):31-35.

Sampson, R.N. and D.L. Adams. 1994. Assessing Forest Ecosystem Health in the Inland West. New York: Food Products Press. 461pp.

Samways, M.J. 1995. Southern hemisphere insects: their variety and the environmental pressures upon them. Pp. 297-320 in: Insects in a Changing Environment, R. Harrington and N.E. Stork, eds. London, U.K.: Academic Press.

SAT (Scientific Analysis Team). 1993. Viability Assessments and Management Considerations for Species Associated with Late-Successional and Old-Growth Forests of the Pacific Northwest. The Report of the Scientific Analysis Team. U.S. Department of Agriculture National Forest System and Forest Service Research. USDA, Forest Service Washington Office.

Schlosser, W.E., K.A. Blatner and R.C. Chapman. 1991. Economics and marketing implications of specific forest products harvest in the coastal Pacific Northwest. Western J. Appl. For. 6(3):67-72.

Schowalter, T.D. 1981. Insect herbivore relationship to the state of the host plant: biotic regulation of ecosystem nutrient cycling through ecological succession. Oilos 37(1):126-130.

Schowalter, T.D. 1985. Adaptions of insects to disturbance. Pp. 235-252 in: The Ecology of Natural Disturbance and Patch Dynamics, S.T.A. Pickett and P.S. White, eds. New York: Academic Press.

Schowalter, T.D. 1986. Ecological strategies of forest insects: the need

for a community level approach to reforestation. New For. 1:57-66.
Schowalter, T.D. 1989. Canopy arthropod community structure and herbivory in old-growth and regenerating forests in western Oregon. Can. J. For. Res. 19(3):318-322.
Schowalter, T.D. 1995. Canopy arthropod communities in relation to forest age and alternative harvest practices in western Oregon. For. Ecol. Manage. 78(1/3):115-126.
Schowalter, T.D. 2000. Insect Ecology: an Ecosystem Approach. San Diego: Academic Press. 483 pp.
Schowalter, T.D. and G.M. Filip, eds. 1993. Beetle-Pathogen Interactions in Conifer Forests. London: Academic Press.
Schowalter, T.D. and P. Turchin. 1993. Southern pine beetle infestation development: interaction between pine and hardwood basal areas. For. Sci. 39(2):201-210.
Schowalter, T.D., W.W. Hargrove and D.A. Crossley, Jr. 1986. Herbivory in forested ecosystems. Annu. Rev. Entomol. 31:177-196.
Schowalter, T.D., D.C. Lightfoot and W.G. Whitford. 1999. Diversity of arthropod responses to host-plant water stress in a desert ecosystem in southern New Mexico. American Midland Naturalist 142(2):281-290.
Schowalter, T.D., Y.L. Zhang and T.E. Sabin. 1998. Decomposition and nutrient dynamics of oak Quercus spp. logs after five years of decomposition. Ecography 21(1):3-10.
Schowalter, T., E. Hansen, R. Molina and Y. Zhang. 1997. Integrating the ecological roles of phytophagous insects, plant pathogens, and mycorrhizae in managed forests. Pp. 171-189 in: Creating a Forestry for the 21st Century, K.A. Kohm and J.F. Franklin, eds. Washington, D.C.: Island Press.
Schowalter, T.D. , B.A. Caldwell, S.E. Carpenter, R.P. Griffiths, M.E. Harmon, E.R. Ingham, R.G. Kelsey, J.D. Lattin and A.R. Moldenke. 1992. Decomposition of fallen trees: effects of initial condition and heterotroph colonization rates. Pp. 373-383 in: Tropical Ecosystems: Ecology and Management, K.P. Singh and J.S. Singh, eds. New Delhi: Wiley Eastern.
Schwartz, M.W., C.A. Brigham, J.D. Hoeksema, K.G. Lyons, M.H. Mills and P.J. van Mantgem. 2000. Linking biodiversity to ecosystem function: Implications for conservation ecology. Oecologia 122(3):297-305.

Seastedt, T.R. 1984. The role of microarthropods in decomposition and mineralization processes. Annu. Rev. Entomol. 29:25-46.

Sedjo, R.A. and D. Botkin. 1997. Using forest plantations to spare natural forests. Environment 39(10):14-20, 30.

Senge, P.M. 1990. The Fifth Discipline: The Art and Practice of the Learning Organization, 1st Ed. New York: Doubleday/ Currency. 424pp.

Seymour, R.S. and M.L. Hunter, Jr. 1992. New Forestry in Eastern Spruce-Fir Forests: Principles and Applications to Maine. Orono, Me: College of Forest Resources, University of Maine.

Shaffer, M.L. 1981. Minimum population sizes for species conservation. BioScience. 31(2):131-134.

Shugart, H.H. and S.W. Seagle. 1985. Modeling forest landscapes and the role of disturbance in ecosystems and communities. Pp. 353-368 in: The Ecology of Natural Disturbance and Patch Dynamics, S.T.A. Pickett and P.S. White, eds. Orlando, Fl: Academic Press.

Simard, S. and E. Vyse. 1994. Paper birch: Weed or crop tree in the interior cedar-hemlock forests of South British Columbia. Pp. 309-316 in: Interior Cedar-Hemlock-White Pine Forests: Ecology and Management, D.M. Baumgartner, J.E. Lotan, and J.R. Tonn, eds. Pullman, WA: Dept of Natural Resource Sciences, Washington State Univ.

Simberloff, D. 1976. Experimental zoogeography of islands: effects of islands size. Ecology 57(4):629-648.

Simberloff, D. 1984. Mass extinction and the destruction of moist tropical forests. Zh. Obshch. Biol. 45(6):767-778.

Simpson, G.G. 1964. Species density of North American recent mammals. Syst. Zool. 13(2):57-73.

Slocombe, D.S. 1993. Implementing ecosystem-based management-development of theory, practice and research for planning and managing a region. BioScience 43(9):612-622.

Smith, D.M. 1962. The Practice of Silviculture, 7th Ed. New York, NY: John Wiley & Sons.

Smith, D.B. 1986. The Practice of Silviculture, 8th Ed. New York, NY: John Wiley & Sons.

Smith, A.T. 1974. The distribution and dispersal of pikas: consequences of insular population structure. Ecology. 55(5):1112-1119.

SNEP (Sierra Nevada Ecosystem Project). 1996a. Status of the Sierra

Nevada. Vol. I. Assessment summaries and management strategies. Final Report to Congess. Wildland Resources Center report No. 36. Centers for Water and Wildland Resources. University of California at Davis. Davis, CA.

SNEP (Sierra Nevada Ecosystem Project). 1996b. Status of the Sierra Nevada, Vol. II. Assessments and scientific basis for management options. Final Report to Congess. Wildland Resources Center Report No. 37. Centers for Water and Wildland Resources. University of California at Davis. Davis, CA. 1528 pp.

Soulé, M.E. 1983. What do we really know about extinction? Pp. 111-124 in: Genetics and Conservation: A Reference for Managing Wild Animal and Plant Populations, C.M. Schonewald-Cox, S.M. Chambers, B. MacBryde and W.L. Thomas, eds. Marlo Park, CA: Benjamin/ Cummings.

Soulé, M.E. 1986. Conservation Biology: The Science of Scarcity and Diversity. Sunderland, MA: Sinauer Associates. 584pp.

Soulé, M.E. 1987. Where do we go from here? Pp. 175-183 in: Viable Population for Conservation. M.E. Soulè, ed. Cambridge, England: Cambridge University Press.

Soulé, M.E and B.A. Wilcox. 1980. Conservation Biology: An Evolutionary-ecological Perspective. Sunderland, MA: Sinauer Associates.

Sousa, W.P. 1984. Intertidal mosaics: patch size, propagule availability, and spatially variable patterns of succession. Ecology 65(6):1918-1935.

Spies, T.A. 1991. Plant species diversity and occurrence in young, mature, and old-growth Douglas -fir stands in western Oregon and Washington. Pp. 111-121 in: Wildlife and Vegetation of Unmanaged Douglas-fir Forests. Gen. Tech. Rep. PNW-GTR 285. Portland, OR: USDA, Forest Service, Pacific Northwest Research Station.

Spies, T.A. and J.F.Franklin. 1991. The structure of natural young, mature and old-growth Douglas-fir forests in Oregon and Washington. Pp. 91-110 in: Wildlife and Vegetation of Unmanaged Douglas-fir Forests. Gen. Techn. Rep. PNW-GTR-285. Portland, OR: USDA, Forest Service, Pacific Northwest Research Station.

Spies, T.A., J.F. Franklin and M. Klopsch. 1990. Canopy gaps in Douglas-fir forests of Cascade Mountains. Can. J. For. Res. 20(5):649-658.

References

Spurr, S.H. and B.V. Barnes. 1973. Forest Ecology, 2nd Ed. New York: Ronald Press. 571pp.

Stanton, M.L. 1983. Spatial patterns in the plant community and their effects upon insect search. Pp. 125-157 in: Herbivorous Insects: Host-seeking Behavior and Mechanisms, S. Ahmad, ed. New York: Academic Press.

Steen, H.K. 1976. The U.S. Forest Service: A History. Seattle, WA: University of Washington Press.

Steffan-Dewenter, I. and T. Tscharntke. 1997. Bee diversity and seed set in fragmented habitats. Pp. 231-234 in: Pollination: from Theory to Practise, 7th International Symp. on Pollination, K.W. Richards, ed. Leiden,The Netherlands: ISHS.

Stickney, P.F. 1986. First Decade Plant Succession Following the Sundance Forest Fire, Northern Idaho. Gen. Tech. Rep. INT-197. Ogden, Utah: USDA, Forest Service, Intermountain Forest and Range Experiment Station. 26pp.

Stone, L. 1897. The artificial propagation of salmon on the Pacific Coast of the United States with notes on the natural history the quinnat salmon. Bull. U.S. Fish Comm. 16:203-235.

Strong, D.R., J.H. Lawton and S.R. Southwood. 1984. Insects on Plants: Community Patterns and Mechanisms. Cambridge, MA.: Harvard University Press.

Swanson, F.J., and J.F. Franklin. 1992. New forestry principles from ecosystem analysis of Pacific Northwest forests. Ecol. Applic. 2(3):262-274.

Swetnam, T.W.. 1993. Fire history and climate change in giant sequoia groves. Science 262(5135):885-889.

Swetnam, T.W. and A.M. Lynch. 1989. A tree-ring reconstruction of western budworm history in the southern Rocky Mountains. For. Sci. 35(4):962-986.

Tappeiner, J.C., D. Huffman, D. Marshall, T.A. Spies, and J.D. Bailey. 1997. Density, ages, and growth rates in old-growth and young-growth forests in coastal Oregon. Can. J. For. Res. 27(5):638-648.

Teensma, D.A., J.T. Rienstra and M.A. Yeiter. 1991. Preliminary Reconstruction and Analysis of Change in Forest Stand Age Classes of the Oregon Coast Range from 1850 to 1940. USDI Bureau of Land Management Technical Note T/N OR-9. Portland, OR: Oregon State University. 9pp plus maps.

Temple, S.A. 1977. Plant-animal mutualism: coevolution with dodo leads to near extinction of plant (Calvaria major). Science 197(4306):885-886.

Terborgh, J. 1974. Preservation of natural diversity; the problem of extinction prone species. Bioscience 24(12):715-722.

Thomas, C.D. 1990. What do real population dynamics tell us about minimum viable population sizes? Conserv. Biol. 4(3):324-327.

Thomas, C.D. and I. Hanski. 1997. Butterfly metapopulations. Pp. 359-386 in: Metapopulation Biology: Ecology, Genetics, and Evolution, I.A. Hanski and M.E. Gilpin, eds. San Diego, CA: Academic Press.

Thomas, J.W., J.L. Parker, R.A.Mowrey, G.M. Hanson and B.J. Bell. 1979. Wildlife Habitats in Managed Forest: The Blue Mountains of Oregon and Washington. Agriculture Handbook 553. Washington D.C.: USDA, Forest Service.

Thomas, J.W., E.D. Forsman, J.B. Lint, E.C. Meslow, B.R. Noon and J. Verner. 1990. A conservation Strategy for the Northern Spotted Owl: A Report of the Interagency Scientific Committee. Portland, OR: USDA Forest Service, USDI Bureau of Land Management, Fish and Wildlife Service, and National Park Service. 427pp.

Thomas, T.W., et al. 1993. Forest Ecosystem Management: An Ecological, Economic, and Social Assessment Report of the Forest Ecosystem Management Assessment Team. US Government Printing Office 793-071.

Tilman, D. 1996. The Benefits of Natural Disasters. Science 273(5281):1518.

Tilman, D. and J.A. Downing. 1994. Biodiversity and stability in grasslands. Nature 367(6461):363-365.

Tilman, D., D. Wedin and J. Knops. 1996. Productivity and sustainability influenced by biodiversity in grassland ecosystems. Nature 379(6567):718-720.

Tilman, D., R.M. May, C.L. Lehman and M.A. Nowak. 1994. Habitat destruction and the extinction debt. Nature 371(1):65-66.

Tobalske, B.W., R.C. Shearer and R.L. Hutto. 1991. Bird Population in Logged Western Larch/ Douglas Fir Forests in Northwestern Montana. Res. Pap. INT-442. Ogden, Utah: USDA, Forest Service, Intermountain Research Station. 12pp.

Torgersen, C.E. 1993. Spatial Variability of Soil Organisms, pH, Moisture, O-horizon Depth, and Temperature in Differentiated

Conifer Stands in an Old-Growth Forest Stand in Western Cascades, Oregon. B.A. Thesis Eugene, OR: University of Oregon. 62pp.

Torgersen, T.R., R.R. Mason and R.G. Campbell. 1990. Predation by birds and ants on two forest insect pests in the Pacific Northwest. Pp. 14-19 in: Studies in Avian Biology. No.13. Avian Foraging: Theory, Methodology and Applications. M.L. Morrison et al, eds. Los Angeles, CA: Cooper Ornithological Society.

Toumey, J.W. and C.F. Korstian. 1937. Foundations of Silviculture Upon An Ecological Basis. New York: Wiley & Sons.

Toweill, D. and P.L. Hanna. 1985. Elk Management Plan, 1986-1990. Boise, Idaho: Idaho Dept. of Fish And Game.

Trappe, J.M. 1962. Fungus associates of ectotrophic mycorrhizae. Bot. Rev. 28:538-606.

Trappe, J.M. and D.L. Luoma. 1992. The ties that bind: fungi in ecosystems. Pp. 17-27 in: The Fungal Community. Its Organization and Role in the Ecosystem, 2nd Ed., G.C. Carroll and D.T. Wicklow, eds. New York: Marcel Dekker.

Trombulak, S.C. and C.A. Frissell. 2000. Review of ecological effects of roads on terrestrial and aquatic communities. Conserv. Biol. 14(1):18-30.

Tuchmann, E.T., K. P. Connaughton, L. E. Freedman, and C. B. Moriwaki. 1996. The Northwest Forest Plan: A Report to the President and Congress. M.H. Brookes, ed. Portland, OR: USDA, Office of Forestry and Economic Assistance.

Turner, M.G., W.W. Hargrove, R.H. Gardner and W.H. Romme. 1994. Effects of fire on landscape heterogeneity in Yellowstone National Park, Wyoming. Journal of Vegetation Science. 5(5):731-742.

Unsworth, J.W. et al. 1991. Elk Management Plan, 1991-1995. Boise: Idaho Dept. Fish & Game. 62pp.

Ure, D.C. and C. Maser. 1982. Mycophagy of red-backed voles in Oregon and Washington USA. Can. J. Zool. 60(12):3307-3315.

U.S. Bureau of the Census. 1978. Census of Manufactures, 1978. Washington, D.C.: Government Printing Office.

U.S. Bureau of the Census. 1990. Census of Population: 1990. Washington, D.C.: Government Printing Office.

U.S. Bureau of the Census. 1991. Statistical Abstract of the United States: 1991. Washington, D.C.: Government Printing Office.

USDOI (U.S. Department of the Interior). 1989. 1985 National Survey of Fishing, Hunting, and Wildlife Associated Recreation. U.S. Department of the Interior, Fish and Wildlife Service. Washington DC: GPO. 167 pp.

USDOI (U.S. Department of the Interior, Fish and Wild Service). 1993. 1991 National Survey of Fishing, Hunting, and Wildlife-Associated Recreation. U.S. Department of the Interior, Fish and Wildlife Service, and U.S. Department of Commerce, Bureau of the Census. Washington DC: GPO. 124 pp. + app.

USDOI (U.S. Department of the Interior) 1998. 1996 national survey of fishing, hunting, and wildlife-associated recreation Washington, D.C.: U.S Deptment of Interior, Fish and Wildlife Service. Available: http://www.census.gov/prod/www/abs/fishing.html.

U.S. Department of Commerce. 1993. Statistical Abstract of the United States. Washington, D.C.: GPO.

USFS (U.S. Department of Agriculture Forest Service). 1963. Timber Trends in Western Oregon and Western Washington. Res. Paper 5. Portland OR :USDA, Forest Service, Pacific Northwest Forest and Range Experiment Station. 154 pp.

USFS (U.S. Department of Agriculture Forest Service). 1969. Douglas-Fir Supply Study. U.S. Forest Service Regional Office. Portland OR: USDA, Forest Service, Pacific Northwest Forest and Range Experiment Station. 53 pp.

USFS (U.S. Department of Agriculture Forest Service). 1976. Timber Harvest Scheduling Issues Study. Review draft. Washington D.C.: USDA, Forest Service. 282 pp.

USFS (U.S. Department of Agriculture Forest Service). 1982. An Analysis of the Timber Situation in the United States: 1952-2030. Forest Resource Report No. 23. Washington D.C.:USDA, Forest Service. 499pp.

USFS (U.S. Department of Agriculture Forest Service). 1988a. The South's Fourth Forest: Alternatives for the Future. Forest Resource Report No. 24. Washington D.C. 512 pp.

USFS (U.S. Department of Agriculture Forest Service). 1988b. An Analysis of the Outdoor Recreation and Wilderness Situation in the United States: 1989-2040, a technical document supporting the 1989 RPA assessment. Draft. H. K. Cordell, project leader. Washington D.C.: USDA, Forest Service .

USFS (U.S. Department of Agriculture Forest Service). 1990. An

Analysis of the Timber Situation in the United States: 1989-2040: Summary. Gen. Tech. Rep. RM-199. Washington D.C.: USDA, Forest Service. 268 pp.

USFS (U.S. Department of Agriculture Forest Service). 1991. Wildlife and Vegetation of Unmanaged Douglas-fir Forests. Gen. Tech. Rep. PNW GTR-285. Portland OR: USDA, Forest Service, Pacific Northwest Research Station.

USFS (U.S. Department of Agriculture Forest Service). 1993a. Eastside Forest Ecosystem Health Assessment. Volume III. Assessment. U.S. Department of Agriculture Forest Service. April 1993.

USFS (U.S. Department of Agriculture Forest Service). 1993b. Region 6 Interim Old Growth Definition for Douglas-Fir Series, Grand Fir/White Fir Series, Interior Douglas Fir Series, Lodgepole Pine Series, Pacific Silver Fir Series, Ponderosa Pine Series, Port-Orford-Cedar and Tanoak(Redwood) Series, Subalpine Fir Series, Western Hemlock Series. Portland, OR: Timber Management Group, USDA, Forest Service. June.

USFS (U.S. Department of Agriculture Forest Service). 1994. Tree Planting in The United States – 1993. U.S. Department of Agriculture Forest Service, State and Private Forestry, Cooperative Forestry. Washington D.C.: Cooperative Forestry. 17 pp.

USFS (U.S. Department of Agriculture Forest Service). 1996. An Integrated Scientific Assessment for Ecosystem Management in the Interior Columbia Basin and Portions of the Klamath and Great Basins. T. M. Quigley, R. W. Haynes, and Russell T. Graham, tech. eds. Gen. Tech. Rep. PNW-GTR-382. Available: www.fs.fed.us/pnw/int-col.htm

USFS/BLM (U.S. Department of Agriculture Forest Service and U.S. Department of the Interior Bureau of Land Management). 1994. Final Supplemental Environmental Impact Statement on Management of Habitat for Late-Successional and Old-Growth Forest Related Species Within the Range of the Northern Spotted Owl. USDA, Forest Service, U.S. Department of the Interior,Bureau of Land Management.

USNRC (U.S. National Resources Committee). 1938. Forest Resources of the Pacific Northwest. Washington, D.C.:GPO

Visser, J.H. 1986. Host odor perception in phytophagous insects. Annu. Rev. Entomol. 31: 121-144.

Vogt, D.J. 1987. Douglas-fir ecosystems in western Washington: Biomass

and production as related to site quality and stand age. Ph.D. dissertation. University of Washington, Seattle.
Vogt, K.A. 1991. Carbon budgets of temperate forest ecosystems. Tree Physiol. 9:69-86.
Vogt, K.A., R.L. Edmonds, and C.C. Grier. 1981. Dynamics of ectomycorrhizae in Abies amabilis stands: the role of Cenococcum graniforme. Holarctic Ecology 4(3):167-173.
Vogt, K.A., D.A. Publicover and D.J. Vogt. 1991. A critique of the role of ectomycorrhizas in forest ecology. Agriculture, Ecosystems and Environment 35(2/3):171-190.
Vogt, K.A., J. Bloomfield, J.F. Ammirati, and S.R. Ammirati. 1992. Sporocarp production by basidiomycetes, with emphasis on forest ecosystems. Pp. 563-581 in: The Fungal Community. Its Organization and Role in the Ecosystem, 2nd Ed. G.C. Carroll and D.T. Wicklow, eds., New York.: Marcel Dekker.
Vogt, K.A., E.E. Moore, D.J. Vogt, M.J. Redlin and R.L. Edmonds. 1983. Conifer fine root and mycorrhizal root biomass within the forest floors of Douglas-fir stands of different ages and site productivities Pseudotsuga menziesii. Can. J. For. Res. 13(3):429-437.
Vogt, K.A., D.J. Vogt, E.E. Moore, B.A. Fatuga, M.R. Redlin and R.L. Edmonds. 1987. Conifer and angiosperm fine-root biomass in relation to stand age and site productivity in Douglas-fir forests. J. Ecol. 75(3):857-870.
Vogt, K.A., D.J. Vogt, H. Asbjornsen, and R.A. Dahlgren. 1995. Roots, nutrients and their relationship to spatial patterns. Plant and Soil. 168-169:113-123.
Vogt, K.A., D.J. Vogt, P.A. Palmiotto, P. Boon, J. O'Hara, and H. Asbjornsen. 1996. Review of root dynamics in forest ecosystems grouped by climate, climatic forest type and species. Plant and Soil 187(2):159-219.
Waddell, K.L., D.D. Oswald, and D.S. Powell. 1989. Forest Statistics of the United States, 1987. USFS, Resource Bull. Pacific Northwest-RB-168. 106 pp.
Waggener, T.R. 1990. Forest, Timber, and Trade: Emerging Canadian and U.S. Relations Under the Free Trade Agreement. Orono, Me: University of Maine Press. 45pp.
Walsh, R.G., D.A. Harpman, J.G. Hof, K.H. John and J.R. McKean. 1989. Long-Run Forecasts of Participation in Fishing, Hunting, and

References

Nonconsumptive Wildlife Recreation. Gen. Tech Rep. SE. Asheville, N.C.: USDA, Forest Service, Southeastern Forest Experiment Station.

Walters, C.J. 1986. Adaptive Management of Renewable Resources. New York: MacMillan.

Waring, G.L. and N.S. Cobb. 1992. The impact of plant stress on herbivore population dynamics. Pp. 167-226 in: Insect-Plant Interactions, Vol. 4., E.A. Bernays, ed. Boca Raton, FL.: CRC Press.

Waring, R.H., and J.F. Franklin. 1979. Evergreen coniferous forests of the Pacific Northwest. Science 204:1380-1386.

Warren, D.D. 1999. Production, Prices, Employment, and Trade in Northwest Forest Industries: Fourth Quarter, 1997. Resource Bulletin Pacific Northwest-RB-230. Portland, OR: USDA, Forest Service, Pacific Northwest Research Station. 130 pp.

Watt, A.S. 1947. Pattern and process in the plant community. J. Ecol. 35(1-2):1-22.

Weatherspoon, C.P. and C.N. Skinner. 1995. An Assessment of factoras associated with damage to tree crowns from the 1987 wildfires in northern California. For. Sci. 41(3):430-451.

Wells, P.V. 1983. Paleobiogeography of montane islands in the Great Basin since the last glaciopluvial. Ecol. Monogr. 53(4):341-382.

Wellner, C.A. 1970. Fire history in the northern Rocky Mountains. Pp. 42-64 in: The Role of Fire in the Intermountain West. Missoula, Montana: University of Montana.

Welsh, H.H. and A.J. Lind. 1988. Old-growth forests and the distribution of the terrestrial herpetofauna. Pp. 439-454 in: Management of Amphibians, Reptiles and Small Mammals in North America, Gen. Tech. Rep. RM-166. Fort Collins, CO.: USDA, Forest Service, Rocky Mountain Forest and Range Experiment Station.

Weyerhaeuser Company. 1994. Habitat Conservation Plan for the Northern Spotted Owl on the Millicoma Tree Farm, Coos and Douglas Counties, Oregon. Weyerhaeuser Company, Millicoma Operations, North Bend, Oregon. November.

Whitford, P.B. 1983. Man and the equilibrium between deciduous forest and grassland Ecosystems, mainly in the Midwestern United States. Geobotany 5:163-172.

Whitlock, C. 1992. The history of Larix occidentalis during the last 20,000 years of environmental change. Pp. 83-90 in: Ecology and Management of Larix Forests: A Look Ahead: Proceedings of an international symposium, Whitefish, Montana. October 5-9, 1992. Gen. Tech. Rep. INT 319. Ogden, UT: USDA, Forest Service, Intermountain Research Station.

Whittaker, R.H. 1960. Vegetation of the Siskiyou Mountains, Oregon and California. Ecol. Monogr. 30(3):279-338.

Whittaker, R.H. 1961. Vegetation history of the Pacific coast states and the "central" significance of the Klamath region. Madrono 16:5-23.

Wickman, B.E., R.R. Mason and H.G. Paul. 1992. Thinning and nitrogen fertilization in a grand fir stand infested with western spruce budworm. II. Tree growth response. For. Sci. 38(2):252-264.

Wiens, J.A. 1997. Metapopulation dynamics and landscape ecology. Pp. 43-62 in: Metapopulation Biology: Ecology, Genetics, and Evolution, I.A. Hanski and M.E. Gilpin, eds. San Diego, CA: Academic Press.

Wild, A. and V.G. Breeze. 1981. Nutrient Uptake in Relation to Growth (Crop Plants, Rhizosphere). Pp. 331-344 in: 30th Proceedings-Easter School in Agricultural Science. University of Nottingham.

Wilderness Society. 1993. The Living Landscape. Vol.2. Pacific Salmon and Federal Lands. A Regional Analysis. Washington, D.C.: Wilderness Society. Bolle Center for Forest Ecosystem Management.

Wilkins, D.A. 1991. The influence of sheathing (ecto-) mycorrhizas of trees on the uptake and toxicity of metals. Agriculture, ecosystems & environment 35(2/3):245-260.

Williams, M. 1992. Americans and Their Forests. Cambridge University Press.

Wilson, E.O. 1988. Biodiversity. Washington, D.C.: National Academy Press.

Wilson, E.O. 1992. The Diversity of Life. Cambridge, Massachusetts: Belknap Press of Harvard University Press.

Wimberly, M.C., T.A. Spies, C.J. Long and C. Whitlock. 2000. Simulating historical variability in the amount of old forests in the Oregon Coast Range. Conserv. Biol. 14(1):167-180.

Winchester, N.N. 1997. The arboreal superhighway: arthropods and landscape dynamics. Can. Entomologist 129(4):595-599.

Wittinger, W.T., W.L. Pengelly, L.L. Irwin, and J.M. Peek. 1977. A 20-year record of shrub succession in logged areas in the cedar-hemlock zone of northern Idaho. Northwest Sci. 51(3):161-171.

Wondolleck, J.M. 1988. Public Lands Conflict and Resolution: Managing National Forest Disputes. New York: Plenum Press.

Yaffee, S.L. 1994. The Wisdom of the Spotted Owl: Policy Lessons for a New Century. Washington, D.C.: Island Press.

Young, J. and J. Newton. 1980. Taming the timber beast. Pp. 21-56 in: Capitalism and Human Obsolescence: Corporate Control vs. Individual Survival in Rural America. J.A. Young and J.M. Newton, eds. Montclair, NJ: Allanheld, Osmun.

GLOSSARY

American marten: *Martes americana*
anadromous salmon: salmon that spawn in freshwaters and migrate to the ocean

bald eagle: *Haliaeetus leucocephalus*
Barrow's goldeneye: *Bucephala islandica*
biodiversity: the number and genetic variability of plant, animals, and microbial species that live in a given location
biomass: broadly, all of the organic material on a given area; in a more narrow sense, burnable vegetation to be used for fuel in a combustion system
black bear: *Ursus americanus canadensis*
black-backed woodpecker: *Picoides arcticus*
boardfoot: unit of measurement for lumber and saw logs; refers to a 12 x 12 x 1 inch board or a segment of a log that will produce boards with these dimensions
boreal owl: *Aegolius funereus*
buffer strip (or zone): strip of land varying in size and shape, intended to preserve or enhance aesthetic values around recreation sites and along roads, trails, or water
brown-headed cowbird: *Molothrus ater*
butt rot: decay or rot characteristically confined to the base or lower bole of a tree

California red-backed vole: *Clethrionomys californicus*

Glossary

canopy: more or less continuous cover of branches and foliage formed collectively by adjacent tree crowns

canyon live oak: *Quercus chrysolepis*

chestnut-backed chickadee: *Parus rufescens*

chinook: *Oncorhynchus tschawytscha*

choker: short length of flexible wire, rope, or chain used to attach logs to a winch line or directly to a tractor

chum: *Oncorhynchus keta*

clearcut: a logging method in which all trees in an area are harvested, regardless of age or size

climax forest: plant community dominated by trees representing the culminating stage of natural succession for that specific locality and environment

climax species: plant species that will remain essentially unchanged in terms of species composition for as long as the site remains undisturbed

coastal cutthroat trout: *Oncorhynchus clarki*

coho: *Oncorhynchus kisutch*

common nighthawk: *Chordeiles minor*

competition: struggle among trees and other vegetation, generally for limited nutrients, light, and water present on a site

conifer: tree that is a gymnosperm, usually evergreen, with cones and needle-shaped or scalelike leaves, producing wood known commercially as softwood

controlled burning: use of fire to destroy logging debris, reduce buildups of dead and fallen timber, control tree diseases, and clear land

crown thinning: removing live growth in a tree crown to admit light, reduce weight, and lessen wind resistance

density: number of trees per unit area of land

dbh: diameter at breast height; measure of a tree at 4.5 feet above ground level

Dolly Varden trout: *Salvelinus malma*

Douglas-fir: *Pseudotsuga menziesii*

Douglas-fir bark beetle: *Dendroctonus pseudotsugae*

down wood: large logs on the forest floor

dwarf mistletoe: *Arceuthobium spp.*

Eastside: area from the Cascade crest east through eastern Washington and Oregon and central Idaho; much drier than the Westside

ectomycorrhizal relationship: a symbiotic relationship between a fungus and the external roots of certain plants

effective population size: the number of breeding adults that would provide the rate of inbreeding observed in a population if mating were random and the sexes were equal in number

Engelmann spruce: *Picea engelmannii*

even-aged stand: stand of trees in which there are only small differences in age among the individual trees

extensive margin: forests previously economically unsuitable for harvesting

extractive jobs: jobs related to timber harvesting and removal

feldfield: rocky soil-less slopes at high elevations

FEMAT: Forest Ecosystem Management Assessment Team, created by President Clinton in 1993

fiber: wood volume produced by a tree or trees that can be converted into wood products, such as lumber, paper, or cardboard; also known as pulpwood

financial rotation: rotation of tree crops determined solely by financial considerations to obtain the highest monetary values over time, in terms of optimum net present value or return on investment

fireweed: *Epilobium angustifolium*

fisher: *Martes pennanti*

flammulated owl: *Otus flammeolus*

flicker: *Colaptes auratus*

forest floor: general term for the surface layer of soil supporting forest vegetation; includes all dead vegetation on the mineral soil surface in the forest as well as litter and unincorporated humus

golden chinquapi: *Castanea pumila*
golden-crowned kinglet: *Regulus satrapa*
golden-crowned sparrow: *Zonotrichia atricapilla*
goldeneye: *Bucephala clangula*
goshawk: *Accipiter gentilis*
grand fir: *Abies grandis*
gray wolf: *Canis lupus*

Glossary

great gray owl: *Strix nebulosa*
grizzly bear: *Ursus arctos horribilis*

hardwood: dicotyledonous trees, usually broad-leaved and deciduous
harvesting: removing merchantable trees
hermit warbler: *Dendroica occidentalis*

Indian Paint fungus: *Echinodontium tinctorium*
intensive margin: forests that already have been supplying wood to markets
ISC: Interagency Scientific Committee

kestrel: *Falco sparverius*

legacies: changes that occur or remain as a result of disturbance, e.g., residual debris, seed banks, advanced regeneration, and surviving plants and animals that influence future change
Lewis' woodpecker: *Melanerpes lewis*
Lobaria oregona: a nitrogen-fixing lichen
lodgepole pine: *Pinus contorta*
loggerhead shrike: *Lanius ludovicianus*
lynx: *Lynx canadensis*

manzanita: *Arctostaphylos*
marbled murrelet: *Brachyramphus marmoratus*
mineralization: the process by which elements present in organic compounds are eventually converted into inorganic forms, ultimately to become available for a new cycle of plant growth
moose: *Alces alces*
mountain bluebird: *Sialia currucoides*
mountain chickadee: *Parus gambeli*
mountain hemlock: *Tsuga mertensiana*
Multiple-use forestry: concept of forest management that combines two or more objectives, such as production of wood or wood-derivative products, forage and browse for domestic livestock, proper environmental conditions for wildlife, landscape effects, protection against floods and erosion, recreation, and protection of water supplies

NFP: Northwest Forests Plan
noble fir: *Abies procera*
northern spotted owl: *Strix occidentalis caurina*

olive-sided flycatcher: *Contopus cooperi*
Option 9: one of the management choices presented to President Clinton by FEMAT in 1993; it has since been revised and termed the Northwest Forests Plan
OSB: oriented strand board
osprey: *Pandion haliaetus*
overstory: tall, mature trees that rise above the shorter immature understory trees

Pacific madrone: *Arbutus menziesii*
Pacific silver fir: *Abies amabillis*
Pacific-slope flycatcher: *Empidonax difficilis*
Pacific yew: *Taxus brevifolia*
patchiness: small islands of regenerating trees scattered through a matrix of older trees
peregrine falcon: *Falco peregrinus*
pileated woodpecker: *Dryocopus pileatus*
pink salmon: *Oncorhynchus gorbuscha*
plantation: forest stand regenerated artificially either by sowing or planting
ponderosa pine: *Pinus ponderosa*
Port Orford cedar root rot: *Phytophthora lateralis*
pulp: mechanically ground or chemically digested wood used in manufacturing paper and allied products
pulpwood: wood cut or prepared primarily for wood pulp and subsequent manufacture into paper, fiberboard, or other products, depending largely on the species cut and the pulping process

rainbow trout: *Oncorhynchus mykiss*
red alder: *Alnus rubra*
red fir: *Abies magnifica*
red-breasted sapsucker: *Sphyrapicus rugar*
red-eyed vireo: *Vireo olivaceus*

Glossary

redwood: *Sequoia sempervirens*
resinosis: exudation of pitch often characateristic of disease
ring rot: rot localized mainly in the springwood of the growth rings, giving a concentric pattern of decayed wood in the cross section of a tree or log
rotation: years between establishment of a stand of timber and the time when it is considered ready for final harvest and regeneration
roundwood products: logs, bolts, or other round sections cut from trees for industrial or consumer use
rufous hummingbird: *Selasphorus rufus*

sapling: young tree less than 4 inches in dbh
SAT: Scientific Analysis Team
scree: loose rocks or stones
silviculture: science of cultivating (such as with growing and tending) forest crops, based on the knowledge of silvics;.more explicitly, the theory and practice of controlling the establishment, composition, constitution, and growth of forests
Sitka spruce: *Picea sitchensis*
snag: a standing dead tree or portion of a tree from which most of the foliage, limbs, etc., have fallen
snowbrush: *Ceanothus velutinus*
sockeye salmon: *Oncorhynchus nerka*
southern pine beetle: *Dendroctonus frontalis*
sporocarp: the fruit cases of certain non-seed-producing plants containing sporangia or spores
spruce budworm: *Choristoneura occidentalis*
stand: a tree community that possesses sufficient uniformity in composition, constitution, age, spatial arrangement, or condition to be distinguishable from adjacent communities
steelhead trout: *Oncorhynchus mykiss*
stocks: salmon: genetically distinct lines
subalpine fir: *Abies lasiocarpa*
substrate: a surface on which a plant or animal grows or is attached
Swainson's thrush: *Catharus ustulatus*

tanoak: *Lithocarpus densiflorus*